DECODING REALITY
THE UNIVERSE AS QUANTUM INFORMATION

DECODING REALITY

The Universe as Quantum Information

VLATKO VEDRAL

University of Oxford, England and
National University of Singapore, Singapore

OXFORD
UNIVERSITY PRESS

OXFORD

UNIVERSITY PRESS

Great Clarendon Street,Oxford OX2 6DP

Oxford University Press is a department of the University of Oxford.
It furthers the University's objective of excellence in research, scholarship,
and education by publishing worldwide in

Oxford New York

Auckland Cape Town Dar es Salaam Hong Kong Karachi
Kuala Lumpur Madrid Melbourne Mexico City Nairobi
New Delhi Shanghai Taipei Toronto

With offices in

Argentina Austria Brazil Chile Czech Republic France Greece
Guatemala Hungary Italy Japan Poland Portugal Singapore
South Korea Switzerland Thailand Turkey Ukraine Vietnam

Oxford is a registered trade mark of Oxford University Press
in the UK and in certain other countries

Published in the United States
by Oxford University Press Inc., New York

British Library Cataloguing in Publication Data

Data available

Library of Congress Cataloging in Publication Data

Library of Congress Control Number: 2009941586

Typeset by SPI Publisher Services, Pondicherry, India
Printed in Great Britain
on acid-free paper by
Clays Ltd., St Ives Plc

ISBN 978-0-19-923769-2

1 3 5 7 9 10 8 6 4 2

Dedicated to my mother – the pain from her absence is only marginally alleviated by my (imperfectly encoded) memories of her.

CONTENTS

Acknowledgements   ix

Prologue   1
1 Creation Ex Nihilo: Something from Nothing   5
2 Information for all Seasons   14

Part One   25
3 Back to Basics: Bits and Pieces   25
4 Digital Romance: Life is a Four-Letter Word   37
5 Murphy's Law: I Knew this Would Happen to Me   57
6 Place Your Bets: In It to Win It   77
7 Social Informatics: Get Connected or Die Tryin'   91

Part Two   111
8 Quantum Schmuntum: Lights, Camera, Action!   116
9 Surfing the Waves: Hyper-Fast Computers   134
10 Children of the Aimless Chance: Randomness
   versus Determinism   152

Part Three   171
11 Sand Reckoning: Whose Information is It, Anyway?   173
12 Destruction ab Toto: Nothing from Something   189

Epilogue   215

Notes   219
Index   227

ACKNOWLEDGEMENTS

There are a number of people who should be thanked and without whom this book would not have existed. I have been influenced by many researchers in the past 15 years of my research career. In alphabetical order those who did most are Janet Anders, Charles Bennett, Sougato Bose, Časlav Brukner, Keith Burnett, David Deutsch, Artur Ekert, Peter Knight, William Wootters, and Anton Zeilinger. They will find much of themselves in the pages of this book.

I would like to thank the departments of physics at the Universities of Oxford and Leeds, UK, and the National University of Singapore who have provided me with a great environment encouraging scientific debate and giving me a better opportunity to communicate science to a wider audience.

Writing is a very solitary exercise and it has been a great pleasure to be able to talk about some aspects of this book to a 'live' audience. Those who attended my various Café Scientifique talks and similar occasions will recognize some material from my talks in here. I am a great believer in taking science out of the University confines and onto the streets. That is where it all started – in the Ancient Greek agorae of Socrates – and that's where it ultimately belongs. I hope that the book preserves some traces of this 'streetwise' style of communicating science.

Luke Rallan's constant help and encouragement as well as reading and commenting on various drafts and ideas are very gratefully acknowledged. He provided much of the stimulus at different stages and without his help this project would never have materialized. I am also grateful to Latha Menon of Oxford University Press and

Kerry McKenzie at the University of Leeds whose extensive comments have improved the manuscript.

I thank various UK and international bodies for funding my research. In particular, the Royal Society and the Wolfson Foundation have been very generous in their support.

My wife Ivona, my sons Mikey and Leo, and my daughter Mia have suffered a great deal in the past two years. They are everything I have – the alpha and the omega – and without them, there would be no reality to decode. I hope to make up for the lost time in the years to come.

PROLOGUE

In autumn 1994, while a final year undergraduate student in London, sorting through my reading material for the forthcoming term, I found three words that would have a profound effect on my future. These three words got me thinking again about questions that I have encountered both in life and in physics. At the time I wasn't exactly sure what they meant, but steadily things started to make sense.

Every child is at some point exposed to various rules (laws and principles) that seem to govern the functioning of the Universe and everything in it. Newton's laws in physics, the photosynthesis cycle in biology, rules of grammar in French, the law of supply and demand in economics, the list just goes on. As a child I remember feeling a little lost and bewildered by these rules that I was required to learn verbatim and just attributed their origin to a magician's trick, something my teacher pulled out of a hat. Further along in life, as our senses and our understanding of the world around us develops, these 'tricks' don't seem quite so intimidating. We are better able to deconstruct them and find that many are not so dissimilar after all. Then at some point, after exploring a sufficient number of these rules across different disciplines, we are in a position to begin speculating on their connection and whether there is a little master book of magic which governs them all. It is this bigger picture that now drives me and many others. Whatever walk of life you come from, the question remains the same: is the reality that we see around us just made up from a seemingly random collection of unrelated rules and events or is there a common underlying thread from which these all derive?

From the dawn of civilization, some of our most inquisitive minds have been pursuing this common thread. By linking together the rubbing of rock on rock or wood on wood we have been able to create fire. By linking together the falling of an apple to the orbits of planets, we have been able to fly to the Moon. By linking together our understanding of molecules with engineering, we have been able to extend human life by successfully performing the vast array of bodily repairs. By linking together our understanding of human nature with communications technology, we have a global market for products and services regardless of the language we speak. Our attempts at understanding and linking different aspects of reality have clearly been very beneficial.

As we continue to further increase our understanding we expect this progress to continue. There is no doubt that any such future development will be based on how well we can interpret new information and connect what we have learned thus far. By continuing to create more connections we can develop more all-encompassing laws which we then in turn use to better understand and affect our reality. In other words, first we break down or decode what we see around us, just to then use this information to construct or encode a better, more well-connected, picture. The big question, of course, is how much can we connect – is it feasible that there is one ultimate law, one master magician's trick, that describes the whole Universe?

Within this discourse, surely the most exciting and fundamental question of all has to be: why is there a reality at all and where does it come from? In other words, before we can even speak about why things are connected, we need to ask ourselves why things exist in the first place. I will argue in this book that the notion of 'information' gives us the answer to both questions. Curiously, this makes information a far more fundamental quantity in the Universe than matter or energy, which is no mean feat in itself. If we look at reality in terms of 'bits of information', it is interesting that both the existence of reality and its inherent connectivity become completely transparent. Irrespective of whether you are a casual reader or a scientific researcher this has extraordinary implications for each and every one of us.

The three words that I read back in autumn 1994, which changed my perspective so markedly, were 'Information is physical'. The three words, in this order, stood out as the title of an amazing chapter in an otherwise obscure book, and over time made me realize that indeed maybe information is the answer. After having spent the last 15 years convincing myself that it is, I now endeavour to spend the next 12 chapters convincing you likewise.

Creation Ex Nihilo:
Something from Nothing

Every civilization in the history of humanity has had its myth of creation. Humans have a deeply rooted and seemingly insatiable desire to understand not only their own origins but also the origins of other things around them. Most if not all of the myths since the dawn of man involve some kind of higher or supernatural beings which are intimately related to the existence and functioning of all things in the Universe. Modern man still holds a multitude of different views of the ultimate origin of the Universe, though a couple of the most well represented religions, Christianity and Islam, maintain that there was a single creator responsible for all that we see around us.

It is a predominant belief in Catholicism, accounting for about one-sixth of humanity, that the Creator achieved full creation of the Universe out of nothing – a belief that goes under the name of *creation ex nihilo*. (To be fair, not all Catholics believe this, but they ought to if they follow the Pope.) Postulating a supernatural being does not really help explain reality since then we only displace the question of the origins of reality to explaining the existence of the supernatural being. To this no religion offers any real answers.

If you think that scientists might have a vastly more insightful understanding of the origin of the Universe compared to that of major religions, then you'd better think again. Admittedly, most scientists are probably atheists (interestingly, more than 95% in the United Kingdom) but this does not necessarily mean that they do not hold some kind of a belief about what the Creation was like and where all this stuff around us comes from. The point is that, under all the postulates and axioms, if you dig far enough, you'll find that they are as stumped as anyone else. So, from the point of view of explaining why there is a reality and where it ultimately comes from, being religious or not makes absolutely no difference – we all end up with the same tricky question.

Every time I read a book on the religious or philosophical outlook of the world I cannot help but recognize many ideas in there as related to some ideas that we have in science. For example, the attitude of 'reductionism' – the fact that we try to reduce everything to a single simple cause – is common to both a religious and scientific way of thinking. While methods of investigation can vary, in the same way that in religion we reduce everything to a common deity, in science we strive towards a unifying theory of everything. In fact this inherent desire to reduce the number of unknowns is prevalent in almost everything we do. Why should this be the case?

Often there are two different reasons given for this natural desire to simplify. First is that we as humans have a very limited imagination and whichever medium we use to understand the world – be it science, religion, philosophy, or art – we will end up exploiting the same limited set of ideas available to us. In other words, even as we begin to describe reality, the ideas that we use are not so different from one another. As an eminent American psychologist, Abraham Maslow, points out, when your only tool is a hammer, every problem looks like a nail. The hammer in our case could be our natural urge to find simple cause and effect relationships. We humans thrive on reducing complexity, finding it more beautiful and more believable to summarize our whole understanding ultimately in terms of one principle (whether it is a single god or a single theory of everything).

It is also important to appreciate that our reality, i.e. our view of the Universe, might be different from the actual Universe itself. We create our reality through our understanding of the Universe and our reality is what is possible based on everything we know. If we heed Maslow's words then we already understand that we are invariably limited, and accept that whatever reality we generate may only ever be an approximation to what the Universe is really like. In this sense, it is somewhat inevitable that as we build and then look to explain our reality, the singular thesis is somehow embedded within this; it is just a very comfortable notion for us to deal with.

Coupled to this, the second reason is that humans are also social beings. Artists, scientists, clergy, and the lay public all exchange ideas with one another, ideas which then feature in each other's work as we try to better understand our Universe, and generate our picture of reality. Notions of beauty and truth in one area inevitably affect ideas in another. With so many compelling arguments, it is perhaps no wonder then that we all follow more or less a similar road.

According to the German philosopher Ludwig Feuerbach, 'Man first unconsciously and involuntarily creates God in his own image, and after this God consciously and voluntarily creates man in his own image'. If we take God as synonymous to reality, then reality and Man's perception of it are, in fact, inseparable. Man creates reality and then uses reality to describe himself. As we strive to reduce complexity, it is again unsurprising that we try to build our reality on the simplest possible causes.

A more optimistic view as to why our ideas somehow converge is that life has evolved in conjunction with the rest of the Universe. We are an embodiment of the same laws that have shaped the Universe and our imagination is intimately correlated to it. Consciously or subconsciously we find that we converge towards these laws. In this view the driver of this convergence, unlike in previous points, is not any limitation on our part to describe the Universe but rather a natural attraction towards the laws that bind it. These views may seem pretty similar, but the main difference is that the latter is more optimistic. Rather than us creating our reality and then only being able to describe the Universe through this reality, it essentially gives us hope that, as we embody the laws describing the Universe we are already on the right track. But will simplicity lie at the end of this journey?

One of the notions that scientists hold in highest esteem is Occam's razor. William of Occam, a fourteenth-century English logician and Franciscan friar, tells us that assumptions should not be multiplied without necessity, or in other words, the simplest explanation is usually best. While you could argue that simplicity is entirely subjective, in Chapter 10 I will show that there is an objective view of simplicity that is universal.

Taking Occam's logic to the extreme would also mean reducing all the explanation about everything in the Universe to a single all-encompassing principle. Imagine how easy this would make our lives; falling in love, the motion of planets, the movements of the stock market, all being explained through this one principle.

But is this really taking Occam to the extreme? Why not try to even get rid of this one principle and deduce everything without any principles? This surely is simpler still and therefore, in line with Occam's logic, must be a better reflection of reality? Deduction without any principles is what the famous American physicist John Wheeler called a 'law without law'. He reasoned that if we can explain laws of physics without invoking any a priori laws of physics, then we would be in a good position to explain everything. It is this view that is the common scientific take on 'creation out of nothing', *creation ex nihilo*.

Gottfried Wilhelm Leibniz, the famous German mathematician and philosopher and one of the inventors of the mathematical technique of calculus, used this logic in his proof of the existence of God. He found it surprising that something, rather than nothing, exists in the Universe, given that nothing is by far the simpler state. The only reason he could find for something to exist at all is that an independent being created that something. This for him was enough evidence to suggest an external influencer – the influencer being God. So even he, like many others, could find no better answer to the *creation ex nihilo* question than postulating a supernatural being.

The trickiness of having a law that explains everything without postulating a law (or some kind of general principle) in the first place was nicely addressed by one of Wheeler's students, Oxford physicist David Deutsch. On this issue Deutsch reasons as follows: 'If there were no all-explanatory physical principle P approachable by the methods of science, this would presumably mean that there exist aspects of the natural world that are fundamentally inaccessible to science.' In other words, if we cannot find an overarching principle, then science cannot explain the Universe and fails in its ultimate objective. Deutsch reasons that any inability to explain the Universe through a single principle P would run directly counter to rationalism and 'to our view of physics as the universal science, which has hitherto been the driving force behind progress in the subject and which we should be extremely reluctant to abandon'.

However, as Deutsch points out, the flip-side of this is also problematic. If there were such an all-explanatory principle P within physics, its origin would be forever insoluble, given that no principle (or law) can explain its own origin or form. It's like asking an air-conditioner 'why are you an air-conditioner and not a chair?'. Clearly the answer lies outside of the air-conditioner, because the air-conditioner itself was just made that way. So, paradoxically, P, the ultimate principle of physics or the law that explains everything, just cannot be. Again its origin must lie outside of physics and hence Wheeler's seemingly self-contradictory expression 'law without law'.

Deutsch's logic shows the fine line we have to walk if we are to try to explain the whole Universe from one single principle. But what exactly is it that this principle is trying to explain? Are we talking about explaining all objects in the Universe, such as chairs and air-conditioners; are we trying to explain social interactions such as falling in love; or are we talking about something more fundamental, like the basic building blocks of matter and their interactions? Surely we need to explain all of this, the origin of all the stuff in the Universe and how it's tied together.

This book will argue that information (and not matter or energy or love) is the building block on which everything is constructed. Information is far more fundamental than matter or energy because it can be successfully applied to both macroscopic interactions, such as economic and social phenomena, and, as I will argue, information can also be used to explain the origin and behaviour of microscopic interactions such as energy and matter.

As pointed out by Deutsch and Wheeler, however, whatever candidate is proposed for the fundamental building block of the Universe, it still needs to explain its 'own' ultimate origin too. In other words, the question of everything from nothing, *creation ex nihilo*, is key. So if, as I claim, information is this common thread, the question of *creation ex nihilo* reduces to explaining how some information arises out of no information. Not only will I show how this is possible, I will also argue that information, in contrast to matter and energy, is the *only* concept that we currently have that can explain its own origin.

So does information also help us find the all-explanatory principle, *P*, discussed by Deutsch? I argue in the third part of this book that when viewing reality in terms of information, this question no longer even makes any sense. We find that the journey itself, in this case the method with which useful information arises, becomes more important than the ultimate destination (the concept of an explanatory physical law). Indeed we question whether there is any ultimate destination at all, or whether, as the Universe evolves, then so does our target, firmly placing the concept of an ultimate physical principle only in our created reality rather than as a necessary construct for the Universe itself.

So what is the important question that we must address? If we agree on information as a natural framework within which to understand our reality, then we should be able to explain all natural phenomena in terms of it. This is the subject of the core of this book, Chapters 3–10. As we go through the chapters, we will see that it is actually the decrease of information that equates to a better understanding of it. Though this might sound odd initially, intuitively we know this to be true – in that when we understand something better we find that we can summarize it within a few basic principles. For example, instead of having 100 different laws to describe the dynamics of a tennis ball being thrown into the air, each law applicable under a different set of conditions, having one law capturing any possible condition is something that we feel gives us a much better understanding. Hence we equate a better understanding of our reality with a compression of the amount of information that it contains.

Conversely, whilst we work tirelessly to reduce the amount of information in our reality, there is a fundamental argument that suggests that the amount of information in the Universe as a whole, if understood correctly, can only ever increase. This is the subject of Chapter 5. This implies that as the Universe reveals more and more to us, our reality of what is and isn't possible consequently grows, leading to more information that then needs to be compressed. The analogy that I often like to give is of a donkey with a carrot hanging at a fixed distance in front of it. As the donkey moves closer to the carrot, thinking he's almost made it, the carrot moves in line with the donkey. The donkey, not realizing that the carrot is attached to it via a stick, continues to try and try, unaware that he is ultimately doomed to failure (it is a donkey after all). While he covers a lot of distance (and gets to know the structure of the carrot intimately), the donkey ultimately fails in his primary objective.

In this sense there is a dichotomy between our desire to compress information (distil our whole understanding of reality into a few encompassing principles) and the natural increase of information in the Universe (the total amount we need to understand). This desire to compress information and the natural increase of information in the Universe may initially seem like independent processes, but as we will explore in much more detail later there may be a connection. As we compress and find all-encompassing principles describing our reality, it is these principles that then indicate how much more information there is in our Universe to find. In the same way that Feuerbach states that 'Man first creates God, and then God creates Man', we can say that we compress information into laws from which we construct our reality, and this reality then tells us how to further compress information.

While some may disagree, I believe this view of reality being defined through information compression is closer to the spirit of science as well as its practice (the so-called scientific method to be discussed in great detail in Chapters 10 and 12). It is also closer to the scientific meaning of information in that information reflects the degree of uncertainty in our knowledge of a system, as will be shown in Chapter 3.

Perhaps the view of the Universe that will be promoted here should more appropriately be called 'annihilation of everything' as opposed to 'creation out of nothing', as ultimately it is compression that we argue defines reality. This will be explained in more detail in Part Three of the book.

Key points

- We present the notion of our reality, which is our under- standing of the Universe and what is and is not possible within it. Our notion of reality continually evolves with our progress.
- We wrestle with the challenge of whether there could be an ultimate law that describes the Universe and the question of how this could arise out of nothingness, *creation ex nihilo*.

This book will argue that information is the underlying thread that connects all phenomena we see around us as well as explaining their origin. Our reality is ultimately made up of information.

Information for all Seasons

Imagine that you arrive late at a party. Everyone is already there, sitting at a big round table. The host invites you to sit down with the others and you realize that they are engaged in what appears to be some kind of a game. The host tells you nothing other than to sit down and join in. Let's say that you quite like playing poker, and you get excited at the prospect of participating, but you quickly realize that this is not poker. Then it dawns on you that you actually have absolutely no idea what is going on. You turn around to consult the host, but he seems to have disappeared. You take a deep breath and keep quiet, not wanting to reveal your ignorance quite so early in the evening, and you quietly continue to observe.

The first thing you notice is that no one is allowed to utter any words, so it's not obvious at all whether this is a game. This seems slightly odd but you think this may be one of the rules of the game and so you play along. You observe that the players are using a common deck of cards, resembling Tarot cards, each card with an elaborate picture on it, such as a warrior killing a lion, or a lady holding two crossed swords. After a while it becomes clear that players take turns to reveal a set of cards, one at a time. As each subsequent card is laid down, adjacent to the previous one, the other players closely observe the card being laid down as well as any body language of the player to further substantiate the meaning of the card.

So it's finally the turn of the player sitting next to you. He puts down a king standing over a dead lion with his sword raised above his head; you think to yourself, 'Is this guy talking about a particular king who killed a lion?', 'Is he talking about royalty in general?', or 'Is this card a metaphor for some kind of personal triumph?'. As you develop your thoughts on this, the next card is placed, which is a red dragon. You think initially that this is some kind of metaphor for danger, but when you look at both cards together, you reason that maybe they represent a Welsh king (the red dragon is the national icon of Wales) or perhaps a powerful person facing danger. The next three cards shown are: two crossed blades, a river, and finally a beggar.

By now it's obvious that they are all trying to convey some kind of message to each other through these cards and their body language. And it's also clear that you probably cannot work out the meaning of this game until you have seen a sufficient number of cards. But you start to ask yourself, what exactly are they trying to convey – what is the main point of this activity? Are they telling their life story, making up a story for each other's entertainment, or perhaps is each combination of cards worth a certain number of 'points'? If it is a game, how do you win it, and if it's not a game, what is the point of it?

A story of this type was imagined by a well-known Italian fiction writer Italo Calvino. The point of his story was that every player is trying to tell others about their life, but only using the images on the cards with a little bit of creative gesticulation and grimacing on the side.

In his book Calvino used this card game as the main metaphor for life. The question is why? Well, it is difficult for me to guess what the writer really wanted to say. Writers are artists, and frequently the point of their work lies precisely in the fact that they are ambiguous and that different people will interpret the same work of art in many different ways. But I am a scientist (as were, incidentally,

Calvino's parents) and I'd like to tell you that what Calvino's card game represents is not all that different from how we generate our understanding of reality.

Calvino's card game is like our dialogue with Nature, in other words, the rest of the Universe. Each of the players at the table represents different aspects of Nature and you are the observer. For example, one player could be economics, one player could be physics, one could be biology, and one could be sociology. Each of the players in turn reveals a little more about their own rules and behaviour as time goes on. Nature, like the players, is silent but reveals its intention through events and the surrounding environment. Unsurprisingly, the language Nature uses to communicate is 'information'. The card game indicates that information comes in discrete units, one card at a time. We cannot divide this card into smaller units. The first message of Calvino's metaphor is therefore that there are basic atoms of information that are universally used. In science, we call these atoms 'bits' or binary digits. We will discuss these 'bits' more precisely in Chapter 3.

The second message of Calvino's story is that any sequence of cards, no matter how transparent its message may seem, has still to be interpreted by the observer (in this case you and the other players). The interpretation may, or may not, be true to what the player intended to convey and may vary widely between different observers, and furthermore the observer himself may have several different views of what he has observed. This is synonymous with the inherent uncertainty we find when we observe Nature and two people may have radically different interpretations.

Interestingly, when it is your turn to play, you become the player and Nature becomes the observer. As you lay down your cards, this reflects back on Nature; there is a duality here – you cannot be at the table without affecting the game. This is the third message of Calvino's story, that in real life you are simultaneously the observer as well as being the player.

The fourth message we can draw from Calvino is that the same card can also mean different things based on which other cards it is drawn with. Regardless of who observes it, each card has its own inherent degree of uncertainty, the same red dragon card can mean danger, fear, or represent the country of Wales, depending on the other cards in the set. Once the whole set of cards is presented, the meaning of each card within this context becomes clearer. Therefore relating Calvino's second and fourth points, these cards, as well as representing bits, depend on who interprets them, as well as the other cards they are drawn with. In this sense, we cannot look at any card individually – they must be considered within the context of the sequence of cards they are drawn with. It's no surprise that this property, in science, goes under the general name of 'contextuality'.

One of the most striking conclusions that follows from this contextuality is that we can never be sure about our interpretation of Nature, given that the next bit of information could falsify our previous view and completely change the essence of the message. In science for example, we could see 1000 experimental results confirming a particular theory, but one subsequent result could completely falsify it and indicate that we have utterly misunderstood the message that Nature is conveying. In Calvino's story, this similarly means that you cannot be sure of the message until the last card in placed on the line. The last card may change the whole point of the story. This is very reminiscent of the Ancient Greek philosopher Socrates' statement that 'no one should be considered happy until they are dead'. You may be happy for most of your life, but until your last breath you can never be sure that you have had a happy life. We will see that the whole edifice of scientific knowledge also rests on this kind of (somewhat brutal) logic.

Analysing Calvino's game a little more also draws some interesting parallels with our observation of Nature. Like the observer in the story, we humans also arrived late to the game. Taking the game as a metaphor for life, if the game has been progressing for 10 years, we only just arrived a couple of minutes ago. Some elements of Nature, such as physics, have been there since the very beginning. So, a huge amount of information has already been conveyed that we haven't yet taken into account as we generate our model of reality.

Calvino takes the players as granted. The scene is already set but Calvino does not tell us why the game started and who invited the players. He leaves this question open, just as it is open in reality. This raises the same issue of where the players come from, and reduces to the challenge of *creation ex nihilo*.

Of course, there is much more to interpreting reality than can be portrayed in any story like Calvino's. It does not give us any concrete details or prescriptions of how exactly we should quantify information and apply it to any given situation, let alone the whole Universe. For example, the arrangement of Tarot cards does not lead us to infer a unique story. How do we decide which story is then more likely than others? Or should we maybe not choose a single story, but combine all the stories into some kind of super-story?

Another crucial aspect missing in Calvino's story, if we use it as an analogue of how Nature presents us with information, is to do with the fact that in Calvino's story, once each card is laid, it cannot be changed. Each card has a definite state (its picture) and, whilst this state may be interpreted differently, it cannot change once it has been laid down. For example, a card showing a red dragon cannot 'magically' change to another card as soon as the next card is drawn, or as soon as it has been observed by someone. As counterintuitive as it may sound, the omission of this interaction between cards, and also between the cards and the players, will be seen to be crucial as we discuss our best physical description of reality, quantum theory, in Part Two of the book.

The reader probably comes to this book with a vague idea of what information is. In everyday parlance information is frequently synonymous with knowledge. We believe we know something when we can talk about it at sufficient length and breadth without being contradicted by any of our listeners. However, although this is the common meaning of the word 'knowledgeable', it is not what a scientist would consider to be knowledge. To a scientist, any knowledge always refers to the knowledge of the future. Hence historians are not scientists – historians make predictions about the past – but science is all about predictions concerning the future. Neils Bohr, one of the grandfathers of quantum theory, jokingly said on this issue that 'It is difficult making predictions, especially about the future'.

Guessing what will happen, means that there is always some risk involved. When we are trying to predict the future, we invariably need to make some leaps of imagination, either because the future is intrinsically uncertain or because we do not have enough information about it. This uncertainty was already explored by Calvino, in that we cannot be sure of the message until the last card is placed on the table. The last card may change the whole point of the story. Unlike in Calvino's story, in which there is a finite set of cards, Nature seemingly lays down cards indefinitely. Unfortunately this means that we have to guess the message that Nature is trying to convey as more and more cards become apparent. As a result, we may be proven wrong by a later card, but this is just a necessary risk inherent in how science works.

Typically, a physicist, when studying an atom say, calculates its properties using pen and paper, or more often these days with the aid of a computer. Then, he goes into the laboratory and makes measurements (these days it is typical for those who calculate and those who measure to be different people, but this need not be so). Finally, the physicist compares measurements with his theory, and if the two coincide to sufficient accuracy he is satisfied that his understanding of the phenomenon is good. If the experiment contradicts the theory – and he is certain that there are no crucial experimental errors involved – then the theory, i.e. our interpretation of the message Nature is trying to convey, must be changed.

This is the basis of the scientific method that has helped us understand various aspects of Nature within a short span of 400 years. It is also this same method that can probably be considered one of the defining features of modern civilization.

Whilst we have thus far been driven by the question of why there is information in the Universe and how it is communicated to us by Nature, our ultimate intention is to show how information describes the reality that we observe. We will do this by following Roger Bacon's dictum of 'analysis and synthesis' and first analyse each of the pillars of reality individually before we synthesize this into an overall unified picture.

Each of the pillars of our reality (players in Calvino's story) will be analysed in terms of how they embody and convey information. While I will be presenting a message from each of these pillars in my own information-centric way, these messages are all well established in the scientific community. The reader may not agree with my ultimate view of encoding reality, but hopefully he or she will find the discussion of the separate pillars valuable in themselves.

The main pillars that we will discuss are:

- **Chapter 4 – Player 1: Biology.** The first major application of information was in biology where genetics developed entirely using the language of information preservation and transmission. Here information is easiest to understand and has a clear and well-defined meaning. Biological information is famed for its endurance, but the underlying principles are in fact universal. We can use them to offer a new framework for running a successful business.

- **Chapter 5 – Player 2: Thermodynamics.** Physics and information have had a long-standing relationship and I use this to talk about the infamous Second Law of thermodynamics. It states that the tendency of the Universe is to decay into chaos. I explain how this is to be understood in terms of information and why it does not contradict the biological preservation of information.

Here I will also use information to present novel insights into the topics of global warming, environmentalism, and offer a new perspective on how to plan your diet.

■ **Chapter 6 – Player 3: Economics.** Having convinced you that biology and physics are all about information, I now claim that human behaviour too is based on the same information-theoretic principles. In particular, betting on random processes, such as in a casino or the stock market, is maximized when we follow these principles. Here we will see how to invest successfully by using laws of information.

■ **Chapter 7 – Player 4: Sociology.** More complex social structures, such as the distribution of cities, the wealth of citizens, and the social order are also seen through the eyes of an information theorist. This chapter is the culmination of the first part of the book, unifying a number of disparate phenomena through one and the same logic. Here I will discuss how to improve your social standing and how racial segregation can occur even within the most ardent group of xenophiles.

■ **Chapter 8 – Player 5: Quantum Physics.** In the second part of the book, I explain that the information in the real world is of a different kind from what it appears to be at first sight. Though still quantified in terms of bits, information is actually far more powerful than what we thought possible. This is because the world is ultimately quantum mechanical. This chapter explains the basics of quantum information which has some bizarre and rather radical features. We will see how to communicate so securely that even the CIA has no chance of eavesdropping on our conversations.

■ **Chapter 9 – Player 6: Computer Science.** This new form of information, based on quantum theory, can be used to compute faster than anything we have seen so far with our PCs (which are in the commonly accepted language of classical computers). Here I explain how hacking into your bank account will only take a

few seconds with a quantum computer and how biological
systems may already be capable of some simple forms of quantum
computation.

■ **Chapter 10 – Player 7: Philosophy.** If the Universe has
quantum information at its core – which is what I start to argue
here – then we revisit the age-old problem of determinism versus
free will. Can we act out of our own accord, or are all our actions
predetermined? Here I will try to convince you that randomness
and determinism do not oppose each other. I present an example
of where they work hand in hand to teleport objects across the
Universe.

Of course, some purists might argue that there is really only one
player in Nature, and that player is physics itself. From the cards that
physics reveals, the hands of all the other players follow. However, in
this book I am arguing that it is the cards that are the most fundamental part of this game. Hence each of the players should be treated
equally even though there may be some repetition in their messages,
e.g. some of what economics reveals to us about human nature has
already been captured by biology.

Once we conclude our analysis we then begin to synthesize these
messages in **Chapters 11 and 12**. The result of this synthesis will
be a reality, encoded through bits of information. Here we will view
the Universe as a big quantum computer, running the biggest
possible computer game to generate our reality. The programmers
are the players in Calvino's card game and their software summa-
rizes everything they've learnt from playing the game. Using the
same logic, we can calculate the amount of information that can be
stored inside any object, even the human brain.

Part Three of the book will also argue that information is the only
appropriate entity on which to base the ultimate theory of everything. Not only does information present a framework in which
gravity can be seen as a mere consequence of quantum theory (inte-
grating quantum theory and gravity is the greatest challenge of

modern physics), but it suggests how information can give rise to the 'law without law' and thereby cut the Gordian knot of *creation ex nihilo*.

Some aspects presented in the final chapters will be speculative or still under discussion in the scientific community, and I will warn the reader when this is the case. However, whilst some of these aspects may turn out to be wrong, I hope that the reader will enjoy the intellectual journey. On this let me finish by quoting the famous eleventh-century Persian poet and astronomer, Omar Khayyam:

> Those who conquered all science and letters,
> And shone as beacons among their betters,
> Did not find the thread of this Tangled Heap,
> Only told a story, then they fell asleep.

Key points

- Calvino's card game forms an effective metaphor for the way that we observe and understand reality.
- Information is the language Nature uses to convey its messages and this information comes in discrete units. We use these units to construct our reality.
- The main players in Calvino's card game represent different aspects of Nature. I have chosen these players to be biology, thermodynamics, economics, sociology, quantum physics, computer science and philosophy.
- Key messages from each of these players will be analysed in an information-centric manner in the chapters to come.
- Synthesis of the key messages from each player will result in our view of how reality is generated or encoded.

PART ONE

Back to Basics: Bits and Pieces

The concept of information is so ubiquitous nowadays that it is simply unavoidable. It has revolutionized the way we perceive the world, and for someone not to know that we live in the information age would make you wonder where they've been for the last 30 years. In this information age we are no longer grappling with steam engines or locomotives; we are now grappling with understanding and improving our information processing abilities – to develop faster computers, more efficient ways to communicate across ever vaster distances, more balanced financial markets, and more efficient societies. A common misconception is that the information age is just technological. Well let me tell you once and for all that it is not! The information age at its heart is about affecting and better understanding just about any process Nature throws at us: physical, biological, sociological, whatever you name it – nothing escapes.

Even though many would accept that we live in the age of information, surprisingly the concept of information itself is still often not well understood. In order to see why this is so, it's perhaps worth reflecting a little on the age that preceded it, the industrial age. Central concepts within the industrial age, which can be said to have begun in the early eighteenth century in the north of England, were work and heat. People have, to date, found these concepts and their applicability much more intuitive and easier to grasp than the equivalent role information plays in the information age. In the industrial age, the useful application of work and heat was largely evident through the resulting machinery, the type of engineering, buildings, ships, trains, etc. It was easy to point your finger and say 'look, this is a sign of the industrial age'.

In Leeds, for example, as I used to take my usual walk down Foundry Street in the area called Holbeck, traces of the industrial revolution were still quite evident. John Marshall's Temple Mills and Matthew Murray's Round Foundry are particularly striking examples; grand imposing buildings demanding respect and appreciation for the hundreds of people who worked in squalid conditions and around the clock to ensure that the country remained well fed, clothed, or transported. Murray is a typical example of an eighteenth-century entrepreneurial industrialist, making his fortune in Leeds by producing locomotives, steam engines, textile machines, and several other machines that all did work by exploiting heat energy.

The process of using energy in the form of heat to produce as much useful work as possible is very simple and intuitive. Feeding hot coal into an engine which then produces steam (heat) to drive the wheels of a train (work) is a process that seems quite easy to grasp from beginning to end. But why can't we say something similar about the information age? To me the concept of information is more widely applicable and even easier to understand than work or heat, so why does it still cause confusion? The answer is that I don't know, but, trust me, by the end of this book, if I have done my job properly, you will find it as easy to identify the role of information in its many guises as Murray found work or heat. As an added bonus, you will find information far more fundamental and widely applicable.

So what do we actually mean when we talk about information? While information is not a difficult concept to understand, it can sometimes lead to confusion given the number of contexts that the word is used in. To make matters worse there is simply not enough accessible material on information around. There has recently been a flurry of books, but many of these are overly technical and do not cater for the non-scientific reader. Being involved with several 'science-communication' initiatives, when asked for accessible introductions to information, much to my frustration my recommendations have been somewhat limited. Fortunately, after 15 years of trying to explain information to myself (and unwittingly to most people I meet) I thought, what better way to burn some calories and midnight oil than to rise to the challenge?

There are two main reasons why the concept of information has not been made more accessible. One is simply the fact that there are many ways in which we could define it. For example, do we define information as a quantity which we can use to do something useful or could we still call it information even if it wasn't of any use to us? Is information objective or is it subjective? For example, would the same message or piece of news carry the same information for two different people? Is information inherently human or can animals also process information? Going even beyond this, is it a good thing to have a lot of information and to be able to process it quickly or can too much information drown you? These questions all add some colour and vigour to the challenge of achieving an agreed and acceptable definition of information.

The second trouble with information is that, once defined in a rigorous manner, it is measured in a way that is not easy to convey without mathematics. You may be very surprised to hear that even scientists balk at the thought of yet another equation. As a result, experts and non-experts alike have so far been avoiding popularizing this concept in a detailed and precise way. Even Stephen Hawking, when writing his bestselling *A Brief History of Time*, was famously advised by his editor that every equation he used would halve the number of copies sold.

In spite of all these challenges, there is an accepted and clear definition of information which is also objective, consistent, and widely applicable. By stripping away all irrelevant details we can distil the essence of what information means within a couple of pages.

Unsurprisingly, we find the basis of our modern concept of information in Ancient Greece. The Ancient Greeks laid the groundwork for its definition when they suggested that the information content of an event somehow depends only on how probable this event really is. Philosophers like Aristotle reasoned that the more surprised we are by an event the more information the event carries. By this logic, having a clear sunny autumn day in England would be a very surprising event, whilst experiencing drizzle randomly throughout this period would not shock anyone. This is because it is very likely, that is, the probability is high, that it will rain in England at any given instant of time. From this we can conclude that less likely events, the ones for which the probability of happening is very small, are those that surprise us more and therefore are the ones that carry more information.

Following this logic, we conclude that information has to be inversely proportional to probability, i.e. events with smaller probability carry more information. In this way, information is reduced to only probabilities and in turn probabilities can be given objective meaning independent of human interpretation or anything else (meaning that whilst you may not like the fact that it rains a lot in England, there is simply nothing you can do to change its probability of occurrence).

There is one more important property of information and together with objectivity it leads to the modern measure of information. Suppose that we are looking at the information in two subsequent but independent events. For example, there is a certain probability that I will go out tonight, say 70%, and also there is a certain probability, say 60%, that I will receive a call on my mobile (this can happen independently of whether I am in or out of my house). So, what is the probability that I will go out *and* receive a call while I am out? Since both events have to happen for this to materialize, the overall chance of this happening is the product of the two probabilities. This comes out to 42% ('70 divided by 100' multiplied by '60 divided by 100').

How about the amount of information in these two independent events? If you are already surprised a little by an event and then another event occurs independently, your total surprise will increase, depending only on the probability of the new event. So the total information in two events should be the sum of the two individual amounts of information, given that they are independent events. Therefore, the formula for information must be a function such that the information of the product of two probabilities is the sum of the information contained in the individual events. Are you still with me? You'll get it, I promise. Amazingly enough it can be shown that there is only one such function that does this job and this function is the logarithm (log for short).

Logarithms were invented by the Scottish mathematician John Napier and have been extremely useful in simplifying long multiplications. The famous French eighteenth-century mathematician, Pierre Simon de Laplace, said about them '…by shortening the labours, [they] doubled the life of the astronomer.' In those days astronomers needed to calculate trajectories of planets and other objects by hand, and this often resulted in whole reams of paper full of calculations. Of course, multiplication is much easier now since we all use calculators and computers, which paradoxically makes logarithms appear outdated and intimidating.

So in summary the modern definition of information is exactly this: the information content of an event is proportional to the log of its inverse probability of occurrence:

$$I = \log \frac{1}{p}.$$

This definition is very powerful because we only need the presence of two conditions to be able to talk about information. One is the existence of events (something needs to be happening), and two is being able to calculate the probabilities of events happening. This is a very minimal requirement which can be recognized in just about anything that we see around us. In biology, for example, an event could be a genetic modification stimulated by the environment. In economics, on the other hand, an event could be a fall in a share price. In quantum physics, an event could be the emission of light by a laser when switched on. No matter what the event is, you can apply information theory to it. This is why I will be able to argue that information underlies every process we see in Nature.

Now that we have our definition of information, which let's face it is not that complicated, we can look at one of the earliest applications of information to solve real-world problems. The story starts with an American engineer, Claude Shannon, back in 1940s New Jersey, at the world-renowned Bell Laboratories.

Even before Shannon, the Bell Laboratories already had a fearsome reputation as a centre of excellence. The Bell Labs were a facility at the height of its powers in the mid-1900s, which went on to win a phenomenal number of awards (including six Nobel Prizes) for contributions to science and engineering and development of a wide range of revolutionary technologies (think: radio astronomy, the transistor, the laser, the UNIX operating system, and the C programming language).

It is no surprise that a major source of pride for the Bell Labs, named after the distinguished inventor Alexander Graham Bell, has always been telephone communications. This is the area Shannon worked in and his role was to investigate how to make communication more secure. For example, when 'Alice' phones 'Bob', she relies on people like Shannon to make sure no unauthorized person can intercept or listen in on her phone-call. At the time this was a hugely pertinent issue, given that America was entering World War II and secrecy had become of paramount importance. After several months of research, Shannon managed to come up with conditions which guaranteed that any communication can be made completely secure against unauthorized eavesdropping. (Interestingly his theory of cryptography is what forms the foundation of modern information security – every time you draw money from an ATM, or make a purchase over the Internet, you have Shannon to thank.)

Through this challenge Shannon became interested in how much people could communicate with one another through a physical system (e.g. a telephone network) in the first place. Shannon thought that perhaps rather than sending only one telephone call down a wire maybe we could send two or three or even maybe more. Of course this wasn't entirely an academic pursuit. When you work for one of the world's largest corporations anything that allows you to squeeze more profit out of your existing infrastructure is ultimately going to be good for your career. Anyway by analysing this question in more detail, Shannon came up with the rigorous definition of information that we discussed earlier (that information is proportional to the log of the probability of an event).

He summarized his findings in a ground-breaking paper in 1948. This paper gave birth to the field of modern information theory and changed the telecommunications landscape forever. The theory that he developed is eponymously referred to as Shannon's information theory.

Shannon imagined two users of a communication channel, Alice and Bob, using a phone line to talk to each other. One thing that Shannon realized was that, in order to analyse the information exchanged between Alice and Bob, he had to be as objective as possible. Shannon didn't care if Alice told Bob 'I love you' or 'I hate you', because from his perspective these two messages have exactly the same length and, ultimately, will earn Bell Labs the same amount of money. Human emotion, as we discussed, is not an objective property of the message, so Shannon discarded it; neither is the specific human language, so this went too. Bell Labs should be making profit no matter whether Alice and Bob are communicating in English, Spanish, or Swahili. The amount of information, in other words, should not depend on the way we choose to express it, but must have a more fundamental representation. Shannon found that the fundamental representation he was looking for had already been developed a century earlier by an English primary school teacher, George Boole.

Boole, whilst working on his grand theory of the *Laws of Thought*, published in 1854, reduced all human thought to just manipulations of zeros and ones. Boole's book began as follows: 'The design of the following treatise is to investigate the fundamental laws of those operations of the mind by which reasoning is performed; to give expression to them in the symbolic language of a Calculus, and upon this foundation to establish the science of Logic and construct its method'. He showed that all such algebraic manipulations that you could want to do can be done just using two numbers, zero and one. A digit that is either a zero or a one is called a binary digit, or a 'bit' for short, and Shannon used the concept of bits to develop his information theory.

As a side-note, interestingly the lack of zero was one of the limitations which prevented the Ancient Greeks from developing a full information theory – zeros just did not exist in Ancient Greece, as it never occurred to them that 'nothing' deserves to be labelled by a number. Zero was, in fact, invented by the Indians, sometime before the birth of Christ, and the Indians communicated this knowledge to the Persians and Arabs in the middle ages, who in turn passed it onto the Europeans. The Europeans, armed with the zero, and some tricks learnt from the Ancient Greeks, now had a more flexible numbering system than the cumbersome Roman numerals. This numbering system was paramount to the progress made in science and mathematics, leading us eventually to the Renaissance, which in turn takes us to the present day. The story of the number zero is entirely fascinating within its own right and is really a topic befitting of an entire book.

Let us return to Shannon's task of optimizing communication between Alice and Bob. Armed with the Boolean universal alphabet, Alice could encode the message 'I love you' into the symbol '1', while the message 'I hate you' could be encoded into the symbol '0'. All we need now is to know the probability with which Alice will send '0' and the probability with which she will send '1'. In other words what is the probability that she loves Bob, and what is the probability that she does not?

Let us say that Bob is quite certain, say with 90% probability, that Alice will send him a '0', indicating that she hates him. Imagine now that he picks up the phone and hears a '1' sent to him via the communication channel. He would be very surprised, given that he attributed a low probability to it (only 10%) and thus this message carries more information. (Of course Alice and Bob do not actually talk in zeros and ones. Alice says 'I love you' or 'I hate you' and it is a device on either side of the phone line that encodes this information into bits and then decodes the bits into the original message of either 'I love you' or 'I hate you'.)

This framework can easily be extended to incorporate more complex messages such as 'Let us meet in front of Nelson's Column in Trafalgar Square in London'. This could be encoded into a string of bits, like the following one: 00110010101000. Naturally we would like to use as few zeros and ones as possible per message as this would make more efficient use of the phone line (i.e. we could pack and send more messages down the channel). The general principle that Shannon deduced is that the less likely messages need to be encoded into longer strings and more likely messages into shorter strings of bits. The rationale behind this is that the messages that we communicate very frequently should be short; otherwise we needlessly waste the phone line capacity. It seems pretty obvious now, doesn't it?

If we consider language as a communication channel, this channel has evolved naturally into a more optimal state. Words we use most, such as 'the', 'of', 'and', 'to', are very short and this is because they have a high probability of occurring. Words that we are least likely to use, in contrast, remain very long as they have a low probability of occurring. In this way we can work out how efficient the English language is in comparison to German, French, or Swahili by seeing how many letters it takes to communicate the most commonly used words and phrases. Interestingly, George Zipf, in 1949, whilst independently analysing languages, came up with a similar argument and found that the frequency of any word is inversely proportional to its rank in the frequency table. Thus the most frequent word will occur approximately twice as often as the second most frequent word, which occurs twice as often as the fourth most frequent word, and so on.

Shannon similarly reasoned that the optimal encoding that maximizes the channel capacity (and the profit for Bell Labs) is to make the length of the message proportional to the 'log of the inverse of the probability of the event occurring'. So the same information measure that we argued for turns out to quantify optimal channel capacity of any communication channel.

There has been a massive amount of work on extending Shannon's information theory directly or indirectly in a variety of disciplines. I jumped on the bandwagon in the late 1990s with my doctoral thesis, which translated how Shannon's information theory could be applied to quantum mechanics. I showed that the basic tenets of Shannon's information theory survive and have much to tell us about our latest model of physics. There is more on this in the second part of this book.

When I finished my thesis, my friends and colleagues gave me as a memento a framed picture of Shannon with all their signatures on the back of it. They knew how much Shannon's work had influenced me and thought that his picture would be an appropriate gift – given that I had spent more time with him than with any of them. In it, Shannon looked like a very thoughtful and distinguished scientist, doing science to better the world around him and to satisfy his own thirst for knowledge. It was a very nice thought.

I want to close this chapter with an amusing story. Shannon did not call his quantity information; he called it *entropy*. What we have so far been introducing as Shannon's information is, in fact, known as Shannon's entropy in the community of engineers, mathematicians, computer scientists, and physicists.

The word entropy appeared once Shannon had derived his 'log of the inverse probability' formula and he approached John von Neumann, a great contemporary Hungarian-born American mathematician, asking for advice on how to name his newly invented quantity. Von Neumann suggested the word 'entropy' to Shannon and an urban myth is that von Neumann did this to simply give Shannon an edge in all scientific debates, since no one really knows what entropy is! Von Neumann is well known for his witty remarks, which makes the urban myth plausible. However the real reason is that Shannon's measure already existed in physics under the name of entropy. The concept of entropy in physics had been developed by the German physicist Rudolf Clausius, some hundred years before Shannon.

Physical entropy has at first sight nothing to do with communications and channel capacity, but it is by no means an accident that the two have the same form. This will be the key to our discussion of the Second Law of thermodynamics and will also offer us insights into economic and social phenomena.

In summary, in order to solve his problem of optimizing channel capacity and to derive his information theory, Shannon stood on the shoulders of many other giants (to borrow Isaac Newton's famous phrase). The giants include the Ancient Greeks, George Boole, John Napier, and John von Neumann. There are no true loners in the world of knowledge – no scientific Clint Eastwoods. But this does not detract at all from Shannon's monumental achievement. Shannon demonstrated intellect, awareness, and drive to piece various ideas together and produce of one the greatest discoveries of the twentieth century.

Key points

- The concept of information is fundamental. It can be given an objective meaning.
- In the information age many of our problems are related to directly optimizing information as opposed to work and heat.
- The basic unit of information is the *bit*, a digit whose value is either zero or one.
- Information is a measure of how surprising something is. Unlikely, low probability events contain a high degree of information. Likewise, high probability events contain very little information.
- If two parties want to communicate efficiently then their messages to one another should be encoded according to Shannon's prescription: unlikely messages should be encoded with many zeros and ones; frequent messages should be given a shorter code.

4

Digital Romance: Life is a Four-Letter Word

Lennon and McCartney were 'spot on' with their rendition of the age-old adage that 'Life goes on'. As well as being one of the catchiest tunes from their 1968 *White Album*, this is also one of the simplest and most profound statements known to us. Underneath its innocuous exterior lies a message that is of fundamental importance. The fact that there was life on Earth almost since Earth was formed clearly demonstrates its robustness, and yet this is all the more impressive given that every day we see how fragile individual living beings appear to be. So the question that comes up again and again in scientific circles is 'how can something that is so imperfect survive for so long?' This was one of the great mysteries of biology.

Interestingly it was a mathematician and not a biologist who made the first major step in answering this question. We met this mathematician earlier, when he advised Shannon to use the word 'entropy' to define his information function – yes, enter once more, John von Neumann. Von Neumann showed how something that was near perfect could be constructed out of imperfect components. This seems a little weird, doesn't it? One would intuitively think that to construct something perfect, each piece would also necessarily have to be perfect. This is one of the central problems for living systems. But how could a mathematician without any access to experimental evidence, and to be fair not much knowledge of biology either, be able to understand life so well? Here is how.

Von Neumann actually asked the question: 'How can we make something of long duration from parts that are very short lived?' Imagine we want to make sure that we write a message down that should exist for the next ten millennia, so that all the future generations can benefit from it. Imagine, for example, that I have discovered the secret to eternal happiness. (I haven't, of course, I am as clueless as you are, though sometimes I think the answer lies somewhere between a cigar and a single malt whiskey.) And imagine that I would like my great-great-great-grandchildren to benefit from my acquired wisdom. How can I ensure that my message lasts for such a long time, when I don't even know anything about the time to come?

The astute reader may begin to see how this is related to a question that Shannon addressed in Chapter 1, that of communication between Alice and Bob. Imagine that, instead of Alice, I am now the encoder of this very important message and the channel, which was the telephone wire in the old picture, is now time. Furthermore, the receiver of the message, previously Bob, is now a generation of people in the distant future to whom we wish to communicate the message. This just illustrates how widely we can interpret the meaning of the communication channel and its users. Just to reinforce the point, you can also think of me – the writer of this book – as Alice; the book as the channel (through which I am communicating my thoughts); and yourself, the reader, as Bob, the receiver of my thoughts.

You may say that the longevity of a message is easy to accomplish. Perhaps we can just make a very solid safe, lock the message inside, and then wait. However your message will only survive for as long as this solid structure does. Natural and man-made disasters, pandemics, disease, or other factors all have a bearing on how long our message survives. The Egyptians thought that pyramids looked pretty solid, but even they have eroded quite a bit over the last six millennia and may perish altogether in the next few millennia. In fact, even our very planet has a chance of being destroyed in the not-so-distant future from a variety of threats (maybe not all external, as you are no doubt aware). Taking all these considerations into account, how could I secure my message to my descendents with a high probability?

This story of transmitting a message through generations is what von Neumann had in mind when he formulated his question. This question also presents a beautiful metaphor for life, whose primary goal is to endure. From the previous discussion it would be quite short-sighted to use a solid immobile structure to contain the message. Such a structure is not necessarily resistant to environmental changes.

What we need is something that is able to deal with its environment and is equipped to respond to whatever is thrown at it. It needs to be able to adapt, move, avoid obstacles and danger whenever it is threatened. But it also needs to be able to deal with its own frailty. Whatever this information carrier is made of, it will always have a finite duration. No battery lasts forever, no heart beats forever.

To illustrate this point, let's consider that conceptually we are able to build a robot that is capable of surviving forever (infinite battery life, non-rusting components, etc.) to carry the information. Furthermore we can assume that this robot could at the same time deal with its complex environment as well as maintain itself when damaged. This could do the trick, but the worries are: how likely are we to make such a device (as we said – no battery lasts forever, no heart beats forever), and, even if we could make it to the best of our abilities, surely its luck would run out at some stage. It is near impossible for the robot to account for every possible damage and environmental influence.

So why produce one robot, why not produce a hundred or a thousand? This seems like a good idea as it will prolong the message, but, ultimately, we still only have a fixed number of robots and the population can only decrease through time. Sooner or later each one of our copies will run out of luck.

But here is where your ingenuity really comes in; you think to yourself: 'Why don't I build a robust robot that can also reproduce itself several times?' It would then give the message to each copy that it creates of itself and the copies would similarly pass the message on to their copies, ad infinitum. So with every generation being capable of carrying the message as well as reproducing itself, we have a fighting chance of preserving the message indefinitely.

And this is in essence what von Neumann explored in his paper on self-reproducing automata. His main contribution was to show how reproduction can be accomplished with robots created out of imperfect parts. This approach was by no means uncontroversial with the scientific establishment during von Neumann's time.

There were two main objections to self-reproduction. First of all, and in the words of von Neumann, 'If an automaton has the ability to construct another one, there must be a decrease in complication as we go from the parent to the construct. That is, if A can produce B, then A in some way must have contained a complete description of B. In this sense, it would therefore seem that a certain degenerating tendency must be expected, some decrease in complexity as one automaton makes another automaton.' This is a pretty damaging objection as it seems to completely contradict everyday experience. Life appears to be getting more and more complex, rather than simplifying into less complex organisms.

Note that what von Neumann calls 'complication' of an automaton is closely related to its information content, i.e. the more complex an automaton, the more bits of information are required to describe it accurately. As a side-note, many different ways of talking about biological complexity have indeed been proposed and we will encounter some of them later in the book.

The second main objection to self-reproduction is related to the previous one, only that now it also seems to contradict logic and not just experience. If A has to make another machine B, it then seems that B needs somehow to be contained within A initially. But imagine that B wants to then reproduce into C. This means that C must have been contained in B, but since B is contained in A, C must also be contained in A. Is your head spinning yet? So, in essence, what we are trying to say is, if we want to make sure that something lasts through hundreds of generations, it would appear that we would have to store all the subsequent copies in the initial copy. If we generalize this to an infinite number of copies, then there is clearly a resource impossibility, given that A would then have to store an infinite amount of information.

This second objection reminds me of the metaphor about a resident of a lunatic asylum, who sets himself the task of painting a complete image of the world in all its minute detail. He needs to start somewhere, so he starts with painting the garden of the lunatic asylum he is in. After some time, just as he becomes very pleased with his depiction of the garden, he realizes that something is missing in his painting. He himself does not appear in the painting! While he painted the whole garden in all its minute detail and complexity, he forgot to include himself – the painter. To correct this he then includes an image of himself, only to discover that the painting is still incomplete. He is still outside of it! He painted himself in the painting, but he – the actual painter who painted the painter – still needs to be incorporated. So he amends this error, and now his painting contains a painter with a canvas containing the painter painting the garden. As he thinks about it more and more, and to his great horror, he realizes that the painting is still incomplete and, worse still, that he can actually never finish the job (being insane doesn't exclude being intelligent, as half of my department can testify to). Unwittingly, the painter has got himself caught up in what mathematicians call an infinite regression.

Referring back to Chapter 1, we saw that Wheeler and Deutsch had the same problem when imagining an ultimate law of Nature that then paints all the other laws. They wanted to have a law that was complete and did not require any laws outside of it to explain it. Similarly, when we compare to the painter, the paradox is that no matter how skilfully this law paints reality, it can never produce a picture that contains everything, because it always fails to take itself into account.

Nature appears to be facing the same challenge when trying to solve the problem of re-creating a living being. One living organism appears to need to store a copy of its successor, who appears to need to store a copy of its successor, and so on and so forth. How can we ever jump out of this infinite sequence? Is life as we know it really a logical impossibility?

Von Neumann was well aware of these objections, which is precisely why he wrote his paper, in order to refute them and show how reproduction is possible both logically and practically, and without reducing the complexity of the new generations. Given that von Neumann didn't have a formal training in biology and was only using the power of abstract thought, it is sometimes simply mind-blowing that he achieved this landmark result.

Von Neumann's key idea is based on the fact that there is a clear separation between different components of the process. If we imagine a message that contains all the instructions for producing copies of an object (say, a house, car, refrigerator, etc.), then a copier, which copies the instructions, along with a constructor, which constructs replicas using the instructions, is essentially all we need to propagate the object indefinitely through time. This is the essence of von Neumann's approach, however for completeness let us spell out all the components needed for the full self-reproduction (known in more modern language as self-replication). The reader may find the next couple of pages quite challenging, but it's well worth the effort to familiarize yourself with details of this landmark result.

Let M be a universal constructor machine, in the sense that it can construct any other object given the appropriate instruction, I. Then let X be a specialized copying machine, which can copy the instruction, I, and insert it into the relevant object M has constructed. For example, let's say that M is a constructor machine within a factory that can construct other complex machines. If we feed M with the instruction to construct a car, then it will construct a car. If we feed M with the instruction to construct a chair it will construct a chair. If we feed M with an instruction to construct itself (i.e. using the instruction which was originally used to construct it) then it will also do likewise and we will end up with two identical machines. Good business you might think. Imagine you need to use this machine in a different part of the factory. Unless you also provide a copy of the instructions, to copy a chair/table, etc. or the machine itself, it's not going to be of much use. So you use a Xerox machine to make a copy of the instructions manual and send it along with the machine.

The process of feeding an instruction, I, and copying the instruction has to be controlled by a mechanism, C. In the factory example the control mechanism would be an administrator feeding instructions to the machine and Xeroxing and including the instruction manual whenever a new copy of the machine is made.

Management will be over the moon, as potentially we can increase production markedly just by replicating the constructor across the factory and then across every factory the company has. If we have 1000 constructor machines then we can make 1000 chairs (cars, widgets) at a time and whenever we need to increase production further we always have the option of asking the constructor to make another copy of itself. To operate the new machine we would require another controller (unless you can convince the previous controller to run both machines simultaneously).

Wait a minute, though: in order to self-replicate, does the new controller know how to use the machine and make copies of the instructions? Probably not, he will have to be trained on both. Perhaps then, looking at this machine in its most general sense as a universal constructor, we should also send instructions on how to make a controller and on how to make the Xerox machine that will be used by the controller. If we could do this it would mean that with sufficient power and materials, this machine would be able to replicate itself indefinitely if required. So let's summarize how this indefinite construction system would work.

What we have to start with are: (i) the universal constructer, M; (ii) the Xerox, X; and (iii) the controller, C. To make this into a perfect self-replicating process we need the full set of instructions not only on how to construct M, but also to construct the controller, C, and the Xerox that the controller uses, X. So the combination of M, C, X, and the instructions on how to construct M, C, and X is then a perfect self-replicating entity, which we will denote E.

The controller takes the instructions on how to construct M, C, and X, and feeds them into M. M then constructs a replica of itself, M', the controller, C', and the Xerox machine, X'. The controller also makes a copy of the instructions, I', using the Xerox machine X. So now we have M', C', X' and I' and this set is ready to be sent out as the fully operational self-replicating entity, E'.

Note that we have avoided the vicious circle we presented before – namely that the first entity needs to contain instructions on every subsequent entity in order to propagate the message indefinitely. The decisive step occurs in constructing an entity containing the universal constructor, the Xerox machine, and the controller, as well as instructions on making all three. This process is legitimate and proper according to the rules of (Boolean) logic. And this is all there is to von Neumann's argument.

Although von Neumann's logic is phrased within the narrow context of self-replication, it clearly avoids the infinite regression that we discussed earlier. Can it therefore also be applied to tackle other similar problems, such as Deutsch and Wheeler's 'law without law'? Could everything in the Universe really arise out of nothing in this same way? We will entertain this possibility in Part Three of the book.

In von Neumann's argument, although there is no logical obstacle to indefinite self-replication, we still have a big practical challenge, namely that we assumed that every part of this process is perfect. However what if there are errors at any stage? What happens if the process of self-replication somehow becomes compromised (for example, the controller forgets to photocopy a page of instructions, the photocopier runs out of toner, or the machine simply breaks down)? The next question therefore is what will happen to subsequent replicas of M that are based on these damaged instructions. It seems like the obvious answer is that the process would have to halt – it simply cannot continue. However, here comes von Neumann's second key insight. He showed that this is not the case and that even imperfect parts can lead to a sustained process of replication. This is achieved through the addition of redundancy, where we create a large number of copies of E. Whilst some of these replicas may be sufficiently affected that they do not pass the controller's quality test, the others will be accepted and will propagate to other parts of the company.

It is also worth highlighting that in reality it is not always the controller that does this quality check; this can also be done by external factors (for example, the environment). We can view proliferation of business franchises in this von Neumann self-replicating framework. Take Starbucks as an example of a successful franchise. The first Starbucks was opened in Seattle in the 1970s. It was obviously sufficiently successful in terms of selling coffee that expansion was a natural option. The challenge was copying the model that had worked for the original Starbucks so well. This has been achieved to an uncanny level of detail and there are now 16,000 near-identical copies of Starbucks in over 30 countries. When you see a Starbucks in Beijing or Athens, you've a fair idea that it's going to look and taste the same as your Starbucks down in the road in New Jersey.

On the other hand, a few Starbucks have closed because they didn't replicate the instruction set accurately – they produced coffee that wasn't faithful to the original brew, or the look and feel was not sufficiently reminiscent to encourage 'punters' to part with their money. Many more have closed despite being perfect replicas of the original. With the latter group, the environment may have forced its closure (e.g. local views being 'anti-Starbucks', local coffee houses being preferred, or even a general trend locally to shift away from coffee houses). For example, 600 closed in 2008 due to environmental factors such as the economic slowdown. Even a perfect replica is not a guarantee of success.

The most successful businesses are those that understand this. They are able to process information continuously from their environment, either through good management or skilful external consultants. This information feeds back into their own set of instructions, along with information on their internal capability to continuously evolve new instruction sets. Unlike living systems, businesses can be geared to modify their instruction sets in very short spaces of time. The speed at which this happens is known as agility. In this sense, agility is the key to business longevity.

Hewlett Packard was founded in a garage in Palo Alto in 1939 by two electrical engineers, William Hewlett and David Packard. They initially concentrated on making electronic test equipment, such as oscilloscopes and thermometers. Later, with the rise of electronics, they moved into semiconductor devices and calculators. In the late 1960s, they saw a niche in the market for minicomputers and they got involved. Today they are known as one of the leaders in personal computing, imaging and printing, and enterprise storage and software. As the environment changed (market demand, information age), HP were able to adapt their instruction set to ride the next wave of technological innovation.

Of course, von Neumann's basic intention was not to explain businesses and their success, or explain how factories could be more productive. He wanted to show that it's possible to build self-replicating robots that could be used to colonize and explore life on other planets. Little did he know that living organisms had already figured this out some three billion years before him!

The holy grail of biology in the 1930s and 1940s was the quest for the structure within a human cell that carries the replicating information so well articulated by von Neumann. This structure was thought to be responsible for the colour of our children's hair, their eyes, their height – it tells us the instruction set of our own operation and every replica that we will produce. There were many people in the race to find this structure including James Watson, Francis Crick, Rosalind Franklin, Maurice Wilkins, Erwin Schrödinger, and Linus Pauling, to name a few. This was an extremely exciting and important time; we were on the verge of something great for humanity – a view into a better understanding of who we are and where we come from.

Ultimately it was a former ornithology student and an ex-physicist who won the race. James Watson and Francis Crick (with help from some of the other noted names) discovered the main carrier of this biological instruction set to be a complex acidic molecule called DNA (deoxyribonucleic acid). DNA contains the instructions on how to produce a similar copy to the organism that carries the DNA, analogous to the instruction set, I, fed to the universal constructor machine, M. Nature is extremely careful with preserving this molecule. It is not just that each of us contains one DNA molecule; in fact almost every cell in every living organism contains DNA. Furthermore each and every DNA molecule is capable enough on its own to reproduce the whole organism (albeit within the right environment). This is an instance of the redundancy that von Neumann discussed. Watson and Crick (and Wilkins) were duly awarded the Nobel Prize in 1962 for physiology and medicine.

From von Neumann's universal constructor we saw that we needed four different components, the universal constructor, M, the Xerox, X, the controller, C, and the set of instructions, I. Together they make up the self-replicating entity, E. Comparing this to life, we can see the cell itself as the self-replicating entity E. Inside the cell there are four different components that enable it to do so:

i)   the protein synthesizer machine, M,
ii) the biological nano-engine (akin to the Xerox copier), X,
iii) enzymes which act as controllers switching the nano-engine on and off, C, and
iv) the DNA information set, I.

To be fair, although the big picture view is now well accepted, there are still many details, e.g. how the nano-engine works, which are still being investigated.

So we see DNA as key to this process, as it contains the blueprint of how each cell operates and replicates. Based on it, the constructor machine within our cells synthesizes amino acids, which in turn make up various proteins and new cells for our bodies. Cell replication is of course an extremely complex process, however in essence it boils down to von Neumann's picture. The crucial step when creating new proteins is how DNA information is faithfully copied (or 'Xeroxed') from one cell to another. We are here only looking at a cell's information carrying capacity because this is the most fundamental signature of durability of life. So how exactly does this Xeroxing process work? What if we run out of toner or paper or make a mistake?

Making a new strand of DNA is like making a new zipper by using the old zipper as a model. A zipper is a little simpler than a strand of DNA because a zipper only has one kind of tooth. Unlike the zipper, DNA has four teeth; A, G, C, and T (these letters represent the names of four different molecules known as bases – adenine, guanine, cytosine, and thymine).

The first thing that the DNA copier does (in most living cells) is to unwind (unzip) a section of the old strand of DNA. It then re-creates a full zip from each of the unwound sections by finding the exact complement for each tooth in the surrounding pool. The rule is that an A tooth can only be matched with a T tooth, and a C tooth can only be matched to a G tooth (and vice versa). This means that whenever the copier spots an A tooth on the strand of DNA, it immediately knows that this should be paired with a T tooth.

The fact that C and G fit together, and A and T fit together, is just like a lock and key mechanism. Some keys are either too big or too small for some locks, but some keys fit perfectly. Because these four combine only in specific pairs, once the unzipped strand of DNA is exposed to the surrounding pool (containing the free teeth) these free teeth float in and line up in the proper order. For example, a free A tooth from the surrounding pool will not generally combine with either C or G. This is the process by which copies of the DNA strand are made.

It is also interesting how Nature uses the idea of redundancy to increase the chances of producing a faithful copy. A group of three bases, such as ATC, is each associated with one amino acid. So given that there are four bases, A,C,T, and G, we have four times four times four, namely 64, possible three-base-long combinations – and hence the possibility of encoding 64 distinct amino acids. However, rather surprisingly, there are only 20 amino acids in total (these 20 make up all living matter, including our bodies) meaning that, rather surprisingly, there is more than one triplet associated with the same amino acid. So, for example, in Nature, ATT, ATC, ATG all encode the amino acid isoleucine, while AGA and AGG both encode arginine.

So what's the point of all this over-encoding? The main advantage is, as before, to help minimize errors when DNA is replicated. So, if instead of ATT the replication process makes a mistake and copies this as ATC (so the last letter has been copied wrongly), this will not even be noticed in the new replica organism since the triplets ATT and ATC simply encode the same protein isoleucine. Nature is leaving little to chance, what an ingenious idea!

Redundancy of this kind – several different sequences of bases to encode one and the same amino acid – is the standard way of error

correcting. This is certainly true for modern computers and communications, but, amazingly, it is also true in the case of all human
cognition.

As an example, here is an extract from an e-mail that hit my inbox
a couple of years back: 'Aoccdrnig to rscheearch at Cmabrigde
Uinervtisy, it deosn't mttaer in waht oredr the ltteers in a wrod are,
the olny iprmoetnt tihng is taht the frist and lsat ltteer be at the rghit
pclae. The rset can be a toatl mses and you can sitll raed it wouthit
porbelm. Tihs is bcuseae the huamn mnid deos not raed ervey lteter
by istlef, but the wrod as a wlohe.' This example just goes to show
that there is a great deal of redundancy in the English language as
well, and not just in the genetic code. If it wasn't for that, my students
would never be able to read most of the comments I scribble over
their work.

Given that my book, like the genetic code, is intended to convey
some information, you may wonder what its redundancy is. The
answer I calculated, using Shannon's entropy from Chapter 1, is 7.4.
This means that instead of writing my message in 200 pages, I probably could have summarized everything I wanted to say in 25 pages.
However, I expect the reader would not have thanked me for this (or
maybe they would?).

You may remember that one of the key points from Chapter 3 was
the importance of using a universal language of discrete binary
digits. You will notice that Nature here also seems to have come up
with a discrete encoding for information. But instead of using two
bases to encode things, as in the Boolean logic we saw earlier, Nature
uses four discrete bases. So why does Nature bother with four bases
when Shannon showed that two is enough to do everything? This is
one of the key questions in biology and we will offer some fascinating speculations on this topic in Chapter 9.

Another very important question to address is why did Nature
choose digital rather than any analogue (non-digital) encoding? In
other words, why did Nature choose just four bases rather than an

infinite continuous set of bases? We don't have a 100% watertight mathematical proof, but we have a pretty good argument to offer as to why the digital encoding is better than any analogue form. There are two reasons in favour of digital encoding: one is the reduced energy overhead to process information, and the other is the increased stability of information processing. Let's look at each of these.

First, let's consider the energy expenditure of information processing. Starting with 10 bits (10 systems with two possible states, zero and one) imagine that it costs a certain unit amount of energy to flip one bit. This is the most elementary information processing we can imagine. Flipping all 10 bits then costs 10 times this unit energy. In order to perform a similar operation in any analogue environment we would have to invest much more energy. In an analogue environment we would need 1024 different states (2 to the power of 10, 2^{10}) to represent 10 bits. This is a necessary characteristic of all analogue systems. As an intuitive insight as to why this is so, think that between the lowest and highest energy state of the analogue encoding we have 1024 units of energy instead of the 10 we had in digital encoding. This huge energy overhead is simply a consequence of the fact that, by definition, the analogue encoding must be treated as a whole (as opposed to bit-by-bit).

The second advantage to Nature's discretization of information in DNA is that of stability. In analogue encoding, again by definition, it is significantly harder to notice errors because analogue encoding is continuous and different states are therefore much harder to distinguish than with digital encoding. So even if Nature started three billion years ago encoding its message as a mix of analogue and digital encodings, you can see why today it is now purely digital. The huge energy expense and much larger sensitivity to errors means that digital encoding was always going to be the natural choice.

But it is not just Nature which chooses this digital route. All modern technology is based on digital principles and, as a result, is far more resilient to errors. Most of the errors now are due to humans or the interface between man and machine. What is reliable is the information processing behind the scenes, deep down inside the operating units of every microchip. The message of all this is therefore pretty clear: when it comes to information processing you should 'be wise and discretize'.

In spite of the care and ingenuity Nature has shown in producing perfect copies of the instruction set *I*, some errors still get through uncorrected. These are called mutations. On average the error rate for DNA replication is around one in a million. This may not seem very high, but consider this: an error rate of 10 in a million could potentially mean that your next child is born a chimpanzee. Most of the mutations tend to be detrimental in that they impair the survival of the constructed living organism, however some mutations may lead to a new and improved form, better suited to its environment. This is the basis of the process of evolution known as 'natural selection'.

Recall that one of the principal objections against any sustained process of reproduction is that complexity of subsequent copies has to decrease. We now know that this is simply not true and we know why. Complexity actually increases on average. The key component in this increase of complexity is the process of natural selection, first clearly identified by Charles Darwin. Natural selection is the process by which you are correlated to your environment because the traits that survive are those that are best suited to that environment. For example, if the whole world is submerged under water, naturally only those organisms which can breathe under water will survive. Moreover, whatever happens subsequent to the flooding, the genetic pool that will propagate will be built from the DNA of these surviving organisms.

Turning this on its head, your DNA can therefore be thought of as a historical record of the environmental changes that your ancestors have lived through (Dawkins describes it as the 'genetic book of the Dead'). Since history is littered with environmental changes of some degree or another, it is clear that any DNA that propagates will only ever get more complex (as it contains more and more information about environmental changes). Note that without randomness this process would not work, given that it is random changes in our DNA that provide the variety on which natural selection operates, selecting mutations that lead to organisms better suited to their environment. As a general theme, we will see that meaningful information necessarily emerges only as an interplay between random events and deterministic selection. Each on its own is insufficient.

Now that we believe we understand the essence of biological information in terms of its durability, it would be useful to use this knowledge to correct genetic errors ourselves. This could potentially eliminate key terminal disabilities and illnesses and possibly improve our overall quality of life. This is the hotly debated and somewhat controversial area of research known as genetic engineering.

The trouble with altering genes artificially is that there is hardly ever a 'one-to-one' correspondence between genes and the traits of an individual. There are no genes that just determine the colour of our eyes. The same set of genes will also determine some of our other features, like height, posture, etc. So if you want to genetically modify the colour of your baby's eyes, then by altering the responsible genes it is likely that you will also affect some other features of your child.

Until we have understood all these genetic inter-relationships it will be very difficult to utilize genetic engineering consistently for any positive purposes. It is not clear to what degree we will be able to understand these inter-relationships to make sure that our changes are always benign. Many hope that this will be the case one day, although Nature may not care one iota for our optimism. Personally, I don't see this as a serious moral dilemma yet, simply because we don't have enough information to consistently affect precise genetic characteristics of humans with a high enough degree of success. However, genetic engineering polarizes opinion between scientists and non-scientists (and even between scientists themselves) like no other issue.

As a postscript to the whole DNA story, I would like to say that Watson and Crick were almost scooped by another Austrian physicist, Erwin Schrödinger. Schrödinger was one of the pioneers of quantum mechanics (we will meet him again in Chapter 9) and after revolutionizing physics, he turned his attention to biology. Schrödinger was no slouch here either, almost deducing the exact mechanism for information reproduction some 10 years before Watson and Crick. His description of reproduction is correct in all its details, apart from Schrödinger thinking that the encoder in the replication must be a crystal (since crystals have a stable and periodic structure seemingly ideal for information carrying and processing). Watson and Crick later proved it not to be a crystal, but in fact an acid (DNA).

It turns out that this story is not really concluded, as some part of the encoding process may be done by some crystal-like structure, which itself carries some extra information. In other words, maybe information is also carried by something other than just DNA – i.e. DNA is not the carrier of all the biological information necessary to reproduce life. We know this because the relevant DNA content of bacteria, frogs, humans, and all living beings is roughly the same. Roughly (only!) 20,000 genes suffice to make any living being. But humans are clearly more complex than bacteria, and so it cannot all be in the DNA. Maybe Schrödinger could still be partly right, but there are also many other competing theories as to where this difference comes from.

On top of this there is the question of where the DNA comes from. Is there a simpler structure from which it has evolved? Going back to Schrödinger's work on the encoding of biological life in crystals, crystals are much simpler in structure than DNA and they grow much more spontaneously in Nature. So maybe, in contest with a number of other theories, they offer some insight into the evolution of DNA itself. Granted that crystals can be created and replicate spontaneously, we still have the question of how the information necessary to reproduce life moved from the crystals to DNA. This idea is not new. It was proposed by Alexander Graham Cairns-Smith some 40 years ago and continues to be a hotly debated area of research in biology.

All these questions are very interesting, but their exact solution is not important here. What is important, and will surely survive any future development in biology, is that at the root of life is the notion of information and the durability of life will depend on how we account for it.

As we saw in the initial chapter on *creation ex nihilo*, the fundamental question is why there is any information in the first place. For the replication of life we saw that we needed four main components, the protein synthesizer machine, M, the DNA Xerox copier X, the enzymes which act as controllers, C, and the DNA information set, I. It already looks very complex, but how did this complexity start from nothing?

Can we turn the question of 'biological information from no information' on its head to perhaps make a bit more sense of it? There is an anthropic principle in science which answers the question of 'Why is the Universe like it is?' with the response that 'If it wasn't, then we wouldn't be here to observe it'. But this sounds like no answer at all. We will be back to wrestle with this issue in the final chapters.

Key points

- Any self-replicating entity needs to have the following components: a universal constructing machine, M, a controller, C, a copier, X, and the set of instructions required to construct these three, I. With these it is possible to then create an entity that self-replicates indefinitely.
- A macromolecule responsible for storing the instructions, I, in living systems is called DNA. DNA has four bases: A, C, T, and G. When DNA replicates inside our cells, each base has a specific pairing partner.
- There is huge redundancy in how bases are combined to form amino acid chains. This is a form of error correction.
- The digital encoding mechanism of DNA ensures that the message gets propagated with high fidelity.
- Random mutations aided by natural selection necessarily lead to an increase in complexity of life.
- The process of creating biological information from no prior biological information is another example of the question of *creation ex nihilo*. Natural selection does not tell us where biological information comes from – it just gives us a frame-work of how it propagates.

Murphy's Law: I Knew this Would Happen to Me

Life now seems so robust that it becomes difficult to imagine how it could ever end. Are we now masters of our own destiny? With the robustness of biological information, combined with deliberate genetic engineering, are we capable of adapting to any environment Nature throws at us? Aside from some exceptional *force majeure* (in which case no payout is guaranteed) is there any condition under which life may end?

One of the most topical and interesting discussions is whether life could run out of energy to function. But how could life ever run out of energy; and what does this actually mean? Are we just talking about the Sun dying or natural resources being depleted? The argument is that however life evolves in the future, it is difficult to imagine how it could run without the basic fuel. So if the Sun does die, we may find ourselves in a bit of pickle. However, in my view this hypothesis is entirely incorrect. At the end of the day, regardless of what happens in the Universe, the total energy is always conserved and it is merely our ability to process this energy that remains in question. Regardless of the Sun dying or natural resources being depleted, the same energy still exists within the Universe, and the challenge would then be to find different ways of harnessing it.

My argument in this chapter is that life paradoxically ends not when it underdoses on fuel, but, more fundamentally, when it overdoses on 'information' (i.e. when it reaches a saturation point and can no longer process any further information). We have all experienced instances where we feel we cannot absorb any more information. The question is: is this fatal?

What would you like the epitaph on your tombstone to read when you die? Usually people do not have a strong desire to inscribe anything grand or meaningful themselves, but their close ones, the family, friends, and relatives, choose to write something down to commemorate their loss. Epitaphs, more frequently than not, contain a very brief description of the person, when they lived, and a statement amounting to the fact that that they will be greatly missed. Cemeteries are ultimately for the living, not the dead.

Perhaps you think similarly and you feel that you should lighten up the mood and atmosphere for the living who come and pay you a tribute. If they bring flowers to your grave, the least you could do is surprise them with an original epitaph. You may therefore be tempted to write something witty and funny that would be entertaining to the crowd gathering around your final resting place. George Bernard Shaw, an English playwright, subscribed to this view, which is why his epitaph reads: 'I knew this would happen to me'.

In physics, the certainty with which Shaw knew of his inevitable death is retold through the Second Law of thermodynamics. Admittedly the Second Law is not as funny as Shaw but it does present this notion in a far more fundamental and widely applicable way.

The Second Law of thermodynamics tells us that in physical terms, a system reaches its death when it reaches its maximum disorder (i.e. it contains as much information as it can handle). This is sometimes (cheerfully) referred to as thermal death, which could really more appropriately be called information overload. This state of maximum disorder is when life effectively becomes a part of the rest of the lifeless Universe. Life no longer has any capacity to evolve and remains entirely at the mercy of the environment.

The Second Law of thermodynamics tells us not only that a system dies when it reaches its maximum disorder, it tells us, startlingly, that *every* physical system must inevitably tend towards its maximum disorder. But life is just another complex physical system, so what is the Second Law telling us about life? It is telling us that even life, one the most robust processes in the Universe, must eventually end – that its death is ultimately inevitable!

The question then is how certain are we about the Second Law. Life is telling us that it can propagate forever, whereas the Second Law is telling us that every physical system must eventually reach its thermal death. So who is right, as these two views seem to stand in direct opposition to each other?

To answer this question it's only right that we now give a little more colour to the Second Law of thermodynamics, one of the most fundamental laws in science. Scientists in fact trust the foundations of the Second Law so much that this is what the great English philosopher Bertrand Russell had to say about it:

> That Man is the product of causes which had no prevision of the end they were achieving; that his origin, his growth, his hopes and fears, his loves and his beliefs, are but the outcome of accidental collocations of atoms; that no fire, no heroism, no intensity of thought and feeling, can preserve an individual life beyond the grave; that all the labours of all the ages, all the devotion, all the inspiration, all the noonday brightness of human genius, are destined to extinction in the vast death of the solar system, and that the whole temple of Man's achievement must inevitably be buried beneath the debris of a universe in ruins – all these things, if not quite beyond dispute, are yet so nearly certain, that no philosophy which rejects them can hope to stand. Only within the scaffolding of these truths, only on the

firm foundation of unyielding despair, can the soul's habitation henceforth be safely built.

What Russell is saying (all in one breath – in a paragraph that contains an eleven-line sentence) is that increase in disorder is so certain that we had better get used to it as fast as possible. No serious philosopher can ignore it. Any belief that we hold, that contradicts the Second Law, does not have much chance of being correct – we really are only deluding ourselves if we think that we can escape its firm clutches.

Incidentally, if you think that Russell's statement sounds a bit depressing, you should read the German philosopher Friedrich Nietzsche. He, in fact, based the whole of his philosophy on the premise that physics implies that life is ultimately pointless, as eventually it must become extinct. The idea of absolute progress (the idea of progress to the point of perfection) must therefore ultimately be an illusion, in direct contrast to the ideas underpinning the evolution of life. Nietzsche thought that this conclusion is so difficult to live with that he needed to introduce the concept of a 'superhuman' – an improved version of the human, able to come to terms with the fact that life cannot achieve absolute progress. Nietzsche, sadly, did not himself have the key attributes of his superhuman – he spent the last 11 years of his life in a lunatic asylum unable to deal with life, disillusioned and alone. A very depressing end to one of history's greatest thinkers.

Scientists, however, are pedantic beings. While Russell's and Nietzsche's arguments hold weight and seem logical from a philosophical perspective, a quantified proof of the Second Law is what scientists require. Only when we are able to quantify something mathematically can we test it reliably to verify or falsify it.

So how could the Second Law be described mathematically? Physics presents one mathematical formulation of the Second Law, based on a quantity known as 'entropy'. This is the quantity von

Neumann was referring to when he suggested to Shannon for him to name his information function similarly (Chapter 3). Entropy is a quantity that measures the disorder of a system and can be applied to any situation in which there are multiple possibilities. Physics presents a mathematical formulation of entropy by looking at all the possible states that the system can occupy. Each one of these states will occur with a certain probability that can be inferred from experiments or from some other principles. The logarithm of these probabilities is then taken and the total entropy of the system is then a direct function of this and tells us its degree of disorder:

$$S = k \log W.$$

Using the concept of entropy physicists recast the Second Law into the principle that the entropy of a closed system always increases. This principle is one of the most fundamental laws in science and has deeply profound and wide-ranging significance for practically everything in the Universe. In fact, you can even think of the Universe itself as a closed system, in which case the Second Law tells us that its entropy is always increasing, i.e. that it is always becoming more and more disordered.

Amazingly, this entropy derived by physicists has the same form as the information-theoretic entropy derived by Shannon. Shannon derived his entropy to convey the amount of information that any communication channel can carry. So in the same sense, maybe we can look at the physicists' concept of entropy as quantifying the information content of a closed system. The Second Law then simply says that the system evolves to the state of maximal information, where no new information can be contained. For those of us using the Internet this will be a very familiar concept. When we are close to the bandwidth of our Internet link, our browser slows down, sometimes dramatically. This really is the information overload that we discussed earlier in this chapter.

When asked in a survey by a popular science magazine *Spiked*, for my opinion on the greatest discovery in physics, I immediately replied that it was Boltzmann's: '$S = k \log W$'. This formula, from one of the founders of modern physics, Ludwig Boltzmann, provides the link between our microscopic and macroscopic understandings of the world. S is the entropy of a system and it signifies how disordered the system is. This is its macroscopic feature. W tells us about the number of its different microscopic states, and k is just a constant Boltzmann derived associating the two. It is Boltzmann's formula that shows us that it is, at least in principle, possible to reduce all our macroscopic knowledge to some basic microscopic physical laws; an attitude and philosophy that is frequently labelled as 'reductionistic'.

Clearly Boltzmann's family thought the same, as this very simple formula for entropy was the epitaph written on his gravestone. Boltzmann discovered the formula in 1870, when he was about 30 years old. He argued that entropy will always grow with time – until it reaches its maximum, and this is exactly another way of stating the Second Law of thermodynamics. At the time this was considered very controversial and Boltzmann met stiff opposition on this and on several other ideas from his closest and most respected colleagues.

Boltzmann, like other great thinkers of our time, suffered a more complicated end than is usual. The pressure of the prevailing scientific establishment clearly was a factor in driving him to suicide. Interestingly, and possibly not accidentally, Nietzsche and Boltzmann are not the only people who suffered a tragic fate after thinking about the consequences of the Second Law. There is also Paul Ehrenfest, who committed suicide, and Robert Mayer, who became insane. Therefore, a disclaimer is possibly appropriate here: should the reader wish to continue reading about the Second Law, they do so at their own risk and I am accepting no liability.

Boltzmann's epitaph is actually not all that different to Shaw's. Only, rather than saying 'I knew this would happen to me',

Boltzmann's says 'I knew my entropy had to reach its maximum sooner or later'. (Admittedly physicists are not as witty as playwrights, but, on the other hand, they probably have a deeper insight into the behaviour of the Universe.)

A very important point is that the Second Law of thermodynamics should not be confused with the energy conservation principle, which is in fact known as the First Law of thermodynamics. The First Law says that energy cannot be created out of nothing. It can only be transformed from one form to another, e.g. from electrical energy to the picture and sound in your TV. It is the First Law that is relevant when we discuss various environmental issues. Our planet has finite stored energy resources, such as coal, oil, and natural gas, which we use to extract useful work, e.g. making plastics, driving our cars, or making our food. The popular concern is that these resources are finite and will eventually be depleted – perhaps forcing us to move to a new planet with fresh resources. But wait a minute! Why should we be moving if energy is conserved? Surely it is just transformed from one form to another, so all we have to do is transform it back into useful energy, so that the cycle can continue.

To answer this, the Second Law enters, and the news is not good! The Second Law tells us that when we convert one form of energy into another we cannot do this with perfect efficiency (i.e. the entropy, the degree of disorder in the process, has to increase). For example, whenever we burn gas to run a car, not all of its energy is converted neatly into the motion of the car; some is lost to less useful effects such as heat and noise. Likewise, we can never draw back together all the energy (the exhaust fumes, motion of car, noise created, etc.) and turn it back into gas in an efficient manner. Some energy just simply has to be lost in this conversion process. This is exactly what the Second Law says: disorder must increase overall and energy must be randomly dissipated to the environment. As a consequence, the environment (e.g. our planet) absorbs this dissipated energy, which manifests itself as a rise in temperature. And so whenever any kind of energy is used we have global warming as a necessary consequence of the Second Law.

In fact, the Second Law is telling us that the only sure-fire way to prevent global warming is simply never to use any energy. Here I am not just talking about avoiding luxuries such as driving cars, using aerosols, or even taking any overseas holidays. Even when you do something as necessary as eating food, you convert it into work, but at the same time, according to the Second Law, this is an inefficient process and you inevitably make things hotter around yourself. Though don't think you can substitute for your home heating system by simply eating more. The increase in temperature is tiny, but the point is that it is still there. One person generates per second as much heat as a typical light bulb (calculating this is one of my favourite exam questions in the first year undergraduate physics that I teach); but when you add up the total contribution from six billion people things do heat up more substantially. To prevent this there is only one way – you must not do anything (though try telling that to your boss). Quit living and there won't be any global warming – at least none as far as man-made causes are involved (now there's a tall order for extreme environmentalists).

The concern for life on this planet is therefore not to run out of energy *per se*, but rather to run out of ways of processing this energy as efficiently as possible. It is interesting that though this is still widely misunderstood at present, Boltzmann, even in 1886, was able to clearly illustrate this point: 'The general struggle for existence of living beings is therefore not a fight for energy, which is plentiful in the form of heat, unfortunately untransformably, in every body. Rather, it is a struggle for entropy that becomes available through the flow of energy from the hot Sun to the cold Earth. To make the fullest use of this energy, the plants spread out the immeasurable areas of their leaves and harness the Sun's energy by a process as yet unexplored, before it sinks down to the temperature level of our

Earth, to drive chemical syntheses of which one has no inkling as yet in our laboratories.'

So what do we actually mean when we talk about a higher entropy environment? Recall that entropy quantifies the randomness of a system. Any physical system is made up of atoms, and more randomness just means that there is more motion for these atoms and more positions for them to occupy within the system. This inevitably induces collisions and a higher temperature within the system. When you feel hot in the room, this is because you are being hit by faster moving atoms transferring their energy to you as they hit you. In a cooler room the converse is true, fewer atoms are moving around at high speed and there is overall less transfer of energy, so you feel cooler. In physics this principle is very important when studying the property of atoms, since it is very hard to see what's going on at high temperatures because the atoms are moving around and rapidly bouncing off one another. So we need to cool the system down in order to slow the atoms and distinguish their behaviour more easily. Of course when we are talking about cooling a system for such studies, it's not just cooling by a few degrees; it's basically to cool it as much as is physically possible (currently to a few billionths of a degree above absolute zero).

We cannot prevent warming up of the Earth according to physics, but what we can and should control is our influence on this warming up process. Our actions will inevitably, though not entirely (contrary to popular belief), affect the rate at which this temperature increases. Ideally we must control this rate, so that by the time the temperature becomes unbearable for us, we have a strategy in place to ensure our survival. This strategy could take the form of evacuating Earth and perhaps colonizing some other part of the Universe. This inevitability of evacuation is not as implausible as one might think. Consider that even a five-degree increase in the overall temperature of the planet would result in a melting of both the polar ice caps, a subsequent increase in sea level, and a significant decrease in land mass not to mention the dramatic changes in weather patterns accompanying this.

The whole point of environmentalism is to make sure that if anything kills us – as a species – it is the First Law, not the Second Law (i.e. that we run out of natural resources before we reach a boiling point). To a physicist, however, the Second Law is far more inevitable and there is no contradiction to it, whereas at least with the First Law we have a fighting chance of finding ways of using different forms of energy. The Second Law can really be a very fast killer (working currently on time-scales of hundreds of years), while with the First Law we could possibly keep going for a little while longer (millions of years). Any hope of surviving indefinitely on this planet is therefore a hope misplaced. We will as a species eventually have to leave; it's purely a case of whether this is sooner or later.

Of course, our planet is a hugely complex system and overall the average temperature has gone both up and down throughout its history. This is because of various 'local' effects. For example, the Earth has been unusually hot in the last 10,000 years, which is apparently one of the main reasons why the human species has been able to develop so effectively, much more than at any other time.

In fact, it will not come as a big surprise that a more developed society consumes more energy. If we take GDP (Gross Domestic Product) per capita as a proxy of how developed a society is, then a recent study (2006) shows a very strong correlation between the degree of development and society's energy consumption. At the top of the list we have the USA closely followed by Japan, Australia, UK, France, Germany, and Canada. All these nations are way above the world average. On the other hand, at the bottom, below the world average are Argentina, Brazil, China, South Africa, and many more developing nations. According to the Second Law, more energy consumption typically also implies a greater entropy increase so therefore this information can be used as a another (better?) indication of one's relative contribution to global warming than just looking at CO_2 emissions. The shining light in this analysis was clearly Japan, being nearly twice as efficient as the US (in terms of having the same degree of development, but only half the energy consumption). So what do we think about a global tax in line with our energy (or entropy) efficiency? Perhaps we are calling for a global market in entropy trading? Well, it might sound crazy but maybe it's not such a bad idea. Anyway you heard it here first folks!

Genetically speaking, humans have existed with pretty much the same makeup for the last 150,000 years. Using genetic information, our origins can be traced to somewhere in Africa, but most of the time before the last 10,000 years, the climate was very harsh, temperatures much lower, and humans were forced to move around quite a lot in order to survive. When you move around in such conditions it is difficult to communicate within peer groups and establish the basic tools to propagate knowledge and culture. In such scenarios it is clear that primitive survival instincts dominate.

In spite of all these fluctuations, according to the Second Law the overall trend is for the planet to tend towards a point of thermal death. The thermal death would happen when the Sun cools down until both the Earth and the other planets all reach the same temperature, though in practice astrophysics tells us that before this happens, the Sun will first blow up into a red giant and destroy the other planets. And after this, it is very difficult to imagine how any kind of life would be possible.

The astute reader may be now reaching a point of profound confusion. It seems that the tendency of entropy in physics is from order (low entropy) to chaos (high entropy). In biology, on the other hand, life generates order and the tendency of living beings is to become less and less chaotic and more ordered (complex). Aren't these two tendencies contradicting one another? Is life trying to violate the Second Law of thermodynamics? Ever since the discovery of the Second Law, this has been a hot topic of discussion.

On the other hand, we could argue that the Second Law and life go hand in hand. We have argued previously that the genetic code has become more complex with evolution and therefore, according to Shannon, requires more bits of information and so has a higher entropy. Thus while entropy in physics increases according to the Second Law, at the same time so does the entropy of the genetic code according to Shannon. So is this increase in entropy in the genetic code related to the Second Law? Does this in fact mean that, far from violating it, life is just a simple consequence of the Second Law of thermodynamics?

Schrödinger was the first to argue convincingly that life maintains itself on low entropy through increasing the entropy of its environment. Of course, this is not in opposition to the fact that the genome may be getting more complex with time. In fact, it may be that you need a more complex genome in order to better utilize the environment and bring yourself to a lower entropy state. Schrödinger expounded his views on life maintaining itself on low entropy in a very beautiful little book entitled *What is Life?* He actually suggested in the very same book that in addition to the energy content of food (for example, the number of calories in a Mars bar) its entropy content should also be displayed. Imagine buying a Mars bar from Tesco with a label telling you that eating this bar will contribute five units of entropy. What does this actually mean? This means that as far as energy is concerned you can survive on Mars bars for a long time but practically you are missing crucial information required to keep your body in a highly ordered (low entropy) state. It is fair to say that the less crucial the information contained in the food, the higher its entropy content.

Of course, high-entropy foods can be formulated with other foods to reduce the overall entropy contribution to your body. This is, of course, the concept of a balanced diet. Again, it is the Second Law that dominates here in that allowing the body to degenerate into a highly disordered state will make you dysfunctional at best and more likely sooner or later will kill you! Morgan Spurlock would agree wholeheartedly with this, after a month of consuming just McDonalds' meals (although, according to the Second Law, he would have achieved the same effect by eating the same energy equivalent of cauliflower, which would still lead to a similarly unbalanced nutritional profile, continuing to increase his entropy. To illustrate that it's not energy but entropy that matters, maybe my first movie should be *Super-Cauliflower Me?*).

All this would lead to the conclusion that the best foods are those for which a certain energy value gives the lowest entropy increase in the body. Now I am going to stick my neck out a little and conjecture that the entropy value of a food is correlated to its completeness, not only in terms of the range of nutrients available but also the bioavailability of the nutrients. So, the Second Law indicates that the kind of food that minimizes your overall entropy production for the same energy cost should clearly be the best way forward. This is another hot and extremely pertinent area of research, and if I find the exact answer, you can read about it in *Professor Vedral's Diet Revolution*.

Interestingly, there is one argument against the Second Law that survived for over a hundred years, before being recently refuted. The argument was presented by James Clerk Maxwell in 1867. He postulated a hypothetical creature, whom we now call Maxwell's demon, which purports to be so clever as to violate the Second Law. Maxwell thought that, while inanimate objects, such as houses, chairs, and mountains, must invariably conform to the Second Law and deteriorate with time, perhaps intelligence can avoid it altogether. The demon manages to convert heat into work without any loss of efficiency. This is in direct opposition to the Second Law. This demon greatly worried scientists at the time, and was threatening to shake the foundations of physics. On the other hand, this would have been great news for humanity, as it was suggesting that work could be extracted from an energy source at zero cost.

Here is Maxwell's ingenious idea. It is so simple, and yet so fundamental, that we will encounter its application in every subsequent
chapter. An example of a non-living physical system that is maximally disordered is the atoms comprising the air in your living
room. They whiz about, bouncing back and forth between the walls
of your room at the staggering speed of 500 metres per second. If
your room is five metres long (probably the average size), then every
atom in it goes 100 times back and forth between the walls of your
room in only one second. Now this, you must admit, is pretty fast.

Atoms move around in a completely disorganized fashion, some
up and some down, some left and some right, and so on. If you
computed their entropy you would find it to be at its maximum, i.e.
the motion of these atoms, given certain energy, could not possibly
be more disorganized.

What Maxwell imagined is a little creature, of molecular size – so
little that we wouldn't be able to see it with the naked eye. But, unlike
molecules, this creature was able to observe the speeds and directions of the travelling atoms. Even more than simply observing, it
would use its intelligence in a very capricious way. The demon would
stand in the middle of the room and every time an atom approached
it, the demon would act like a traffic policeman and do the following;
if the demon, using its speedometer, observed that the atom was
moving quickly then it would direct it to one side of the room.
However, if the atom was moving slowly, it would direct it to the
other side of the room. Any redirection would be done without
affecting the speed of the atoms.

This process would separate slow moving molecules to one side
and fast moving molecules to the other side of the room. In other
words, it would be able to introduce some order into what was
initially a very disordered system – without costing any energy.
Maxwell showed that all the demon really needs to do is measure
and think, whilst all the demon's other actions could happen without
dissipating any energy.

Creating order out of disorder is precisely what the Second Law tells us it isn't possible to do! Disorder simply must always increase and persist. For you, the demon's ordering of the atoms in the room would effectively manifest itself in the room being of unequal temperature. The side with slower atoms would be cooler than the one with faster atoms, and when we have a temperature difference then we also have a capacity to do work, which effectively implies a free lunch (energy for no cost).

Maxwell was therefore very worried. While lifeless physical objects surely conform to the Second Law, it seemed to him that life, intelligent life in particular, could easily violate it.

The Second Law is only valid for isolated systems, those that do not interact with other systems. However, no living being is an island. We exchange energy and matter with our environment, and this was precisely where Schrödinger's point comes in. Living systems stay alive by sucking out energy and low entropy (information) from their environments. So, a living system can and does reduce its disorder, but this is always at the expense of increasing the disorder elsewhere in its environment (hence our discussion on global warming).

Even the Earth itself is not an isolated system – it receives energy from the Sun in the form of sunlight. Evolution of life may on its own decrease disorder, but the Earth is getting hotter as a consequence, and the Sun will eventually cool down. Once the Earth and the Sun reach the same temperature, i.e. thermal death, life will no longer be possible, even in principle, let alone in practice. And this now shows us why life needs to be very creative to survive. Only von Neumann's 'robots', that are capable of reproducing, exploiting, and ultimately fleeing their environments, will endure in this Universe.

Let's come back to Maxwell now. What about his demon – what do we make of it? Can such a being exist given all the above discussion? Can anything reduce the total entropy? Not just its own disorder, but the disorder of the total Universe including itself?

The Hungarian physicist Leo Szilard – one of the people who helped create the atomic bomb during World War II – realized that the key insight into why the demon does not violate the Second Law lies in using the concept of information. What he realized in very simple terms is that all the demon had to do was to process some very simple form of information. Maxwell's demon can, in other words, be demoted from a 'supernatural' being to an ordinary computer. The important point in Szilard's observation is that a demon could be made into a physical system – there just isn't anything important about intelligence, and the demon could simply and blindly be following a computer program code (note that Szilard was discussing these concepts 10 years before computers were even invented). The key implication of the demon being a physical system is that, as a result of the Second Law, it has to heat up for one reason or another. Heating up is simply a manifestation of its increase in entropy resulting from the demon gaining the information. Szilard thought that it was the measuring of atoms' speed that required the demon to do the work and therefore heat up. And so, concluded Szilard, the demon cannot exist, even in principle.

Many people have subsequently reached the same conclusion. The most interesting step was taken in the 1960s by an American physicist, Rolf Landauer of IBM. Following closely in the footsteps of Szilard, he concluded that not just demons, but in fact any computer has to heat up as it functions. Like Maxwell's demon, a computer processes information as it runs and any information processing – so Landauer argued – must lead to wasting of heat. So the heating of computers is as certain as the Second Law of thermodynamics.

We already know this directly from experience. If you keep your PC or laptop on for a long time, you notice that it becomes hotter. This heating effect is ultimately a necessary consequence of the way that computers process information. We can even do a small calculation to estimate how hot things can get when computers calculate. A computer can perform say a billion calculations in a second.

Current computers generate of the order of one million basic units of heat per calculation. Therefore if we assume some reasonable properties of your room (and multiply this by the product of Boltzmann's constant and the temperature), all this amounts to a few degrees increase of temperature in a day. So computers are very effective heat radiators (some perhaps more effective at radiating heat than computing).

But is this temperature increase absolutely necessary in any kind of information processing? Can we not have smooth, frictionless information processors that just run without dissipation, and in the most environmentally friendly way? The answer to this question is very interesting.

If memory is properly configured it can keep track of all the information processing without any increase in heat and disorder. However, when the memory is full, then in order to continue to operate – rather than suffering an information overload – it needs to reset itself or delete some of the information. You may think 'so what's the problem, just press the delete key', but the fact that we cannot just press a delete key and forget about the information that we have just deleted is ultimately what proves that 'information is indeed physical'.

When we 'delete' information all we actually do is displace this unwanted information to the environment, i.e. we create disorder in the environment. The dumping of this information into the environment, by definition, results in an increase in the entropy of the environment and therefore an increase in its temperature. It is for this reason that computers are fitted with little fans in order to remove the heat generated by the components as information is continually erased.

This logic was used by Charles Bennett, an IBM colleague of Landauer, to finally exorcise Maxwell's demon. So even if the demon processes information for measuring speeds, it must have a finite memory and hence must eventually delete this information to continue. It is this deletion of information from the demon's memory that increases the information in the environment by at least the amount of work that has been done by the demon. The demon, even in a perfect setup, therefore cannot violate the Second Law.

The main message from Landauer and Bennett's work is that information, rather than being an abstract notion, is entirely a physical quantity. In this sense it is at least on an equal footing with work and energy. So now we have brought information up to the important level of energy and matter. But, as I promised, I will show that information is even more fundamental than this. I first caught a glimpse of this realization whilst I was an undergraduate in London. The three words 'information is physical' stimulated a new perspective and would profoundly affect how my own research would evolve.

Take the human brain as an example of an information processing device. The information inside our heads and the speed at which it is processed still exceeds the capacity of any computer (note that this won't be true in a few years time if the current trend continues!). We have some ten billion neurons in the brain. They are responsible for moving electrical impulses around our head and our body. Whenever someone touches you, neurons generate a signal at the point where the touch happened and this signal then travels along a network of nerves up to your brain (and sometimes other parts of your body).

Let us assume that every nerve cell or neuron can hold one bit of information, namely a zero or a one. This bit is encoded as the presence or absence of an electrical signal in neurons, i.e. when there is an electrical signal the brain detects a one, and when there is no electrical signal the brain detects a zero. This is probably an oversimplification but let's keep the story simple. So, our brain can hold ten billion bits of information. Once we have used up all the bits of memory in our head, in order to be able to record anything further you would first have to delete some information i.e. you would have to forget. This will make your environment hotter, since you have to dump some memories elsewhere. So it is the forgetting (not the forgiving) that requires energy.

The reader may be wondering how much order we could create by using up all the memory our brain allows. The result is very surprising; we would generate a million times less order than would be required to cool down a normal bottle of water by one degree. This is something your fridge at home could do in a matter of seconds.

So why does a home PC generate so much heat in the environment when it only has a fraction of our brain's computational power? This is because our brain is far more efficient as an information processor and only dumps to the environment when absolutely needed. While a computer uses one million units of energy per calculation our brain uses only one hundred. To be fair to computers they have only existed as such for the last 60 years, whereas life has been working at this game for the last three and a half billion years.

In Chapter 4 we argued that at the core of life is the information processing embedded in DNA replication. The central calculation that the DNA performs is matching base pairs and each of these matches costs approximately a hundred units of energy. To re-create a whole new strand of DNA then requires around one billion units of energy. This information processing increases the temperature of the environment.

Each of these units of energy we are talking about directly depends on the temperature. What if information processing were altogether done at the absolute zero of temperature? (This is approximately −273 degrees Celsius.) Would computers still heat up? Interestingly, thermodynamics tells us that, if we could process information at absolute zero, then there would actually be no heat dissipation during computation. But, guess what? Physics prohibits any object from reaching this mystical temperature. This is known as the Third Law of thermodynamics. So, there is no way to win when one plays with

Nature – there just isn't any free lunch. Thanks to information, common sense wisdom has been scientifically vindicated after all: 'no pain, no gain!'

Key points

- Physical entropy, which describes how disordered a system is, tends to increase with time. This is known as the Second Law of thermodynamics.
- Physical entropy is closely related to Shannon's entropy, and is in fact an instance of it.
- The increasing complexity of life is driven by the overall increase in disorder in the Universe.
- Systems which exploit disorder are called Maxwell's demons. All living systems are Maxwell's demons, but so are some non-living things, e.g. computers. Computers perform well in specific calculation tasks and use energy, in the form of electricity, to do work and produce heat. All demons work this way.
- As a result the environment which demons populate increases in temperature. All life and therefore, all cars, and computers contribute to global warming, which is necessary according to the Second Law.
- Information, far from being an abstract concept, has been shown to have a very real physical representation.

Place Your Bets: In It to Win It

Until now we have discussed how life propagates and how life eventually ends; but I guess what most of us are preoccupied with is 'what we do in between.' In this chapter, I would like to stay in-between these two extremes and enjoy the moment. What more could we ask for? Excitement is what I, for one, would like to have.

Whilst the concept of excitement may be subjective, most would agree that some modicum of risk comes as a given. It is much harder to get excited by certainty (let's face it, we all find certainty boring). Let us instead choose life and discuss the various ways to make it more exciting.

It's 1962 Las Vegas, the city of dreams. Millions are made and lost every minute of every day. The city is littered with dreams of rookies making their way across the Nevada desert with borrowed money to chance their arm. Perhaps he will come back a millionaire or perhaps he will come back with his tail between his legs. But this day is different. Today a new cowboy is in town. He enters one of the casinos, the music is going, the cameras are on him, and the wine and the girls are on tap. He looks around, spots the blackjack table and makes a beeline straight for it. When the sexiest game in town is poker – why is this guy spending all his time on the blackjack table? He has a strategy, he thinks, that will beat the dealer. In his pocket, he has $10,000 to play with (in 1962 not an insignificant amount – worth around a cool quarter of a million dollars today), so this guy clearly means business.

He starts to play the game like any rookie, placing small bets, quite innocuous, but as the game wears on, whilst others were leaving the table, this guy is still going. Slowly but surely his strategy seemed to be working. Of course, no casino likes winners, and is particularly wary of those that go about their business with such ruthless efficiency in such a cool and methodical manner. While casino security were keeping a close eye on him, they just couldn't see where the angle was; why was this guy consistently winning? Why couldn't he just be like everyone else and lose? So they let him play for another couple of hours, before they decided that enough was enough and escorted him out of the building. It's an interesting fact that any casino can kick you out just because you are starting to win more than their statistical expectation (which is that on the whole you should be losing!). Over the weekend this guy played virtually every casino in Las Vegas, and by the end had been expelled from them all. In total he walked away from Vegas with a cool $21,000 in winnings (more than doubling his initial investment).

His name was Edward Thorp and, as it turn out, he was a professor of Mathematics at MIT. He later wrote a book on his experiences and betting strategies, *Beat the Dealer*, which subsequently became a best seller, selling over 700,000 copies. So what was his strategy? (And by the way, who the hell lends a quarter million dollars to a university professor to gamble with anyway?)

Edward Thorp was also excited by taking risks. However he knew that if he played in a certain way his risk taking was limited, and that he would be able to get a good return in the long term. So what exactly is the right amount of risk taking? This amount, too – believe it or not – is governed by Shannon's laws of information, providing that you're in it to win it.

Edward Thorp was familiar with the research of Shannon and Robert Kelly, a colleague of Shannon's at Bell Labs. Kelly's paper came some 10 years after Shannon's and was the first to apply Shannon's ideas to betting. This was the first piece of work extending Shannon's information theory. It is rather surprising that in 10 years there was no other extension. The reason for this was apparently not that it took a long time to accept Shannon. On the contrary, his paper was an overnight success. The point was, instead, that Shannon's results were so complete and perfectly argued that no-one had anything to add to it! So, what did Kelly manage to do?

Remember that the information gain is very large when something unlikely happens and this is what leads to the greatest surprise and (possibly) excitement. Well, the best way to get excited is to enter a casino and place a bet on an unlikely thing, like throwing snake eyes (that is, two dice each coming up with a number one in a single throw). This has a 1 in 36 chance of happening (because there are six numbers on each die and six times six is 36). This event is unlikely and it is precisely because of it that you would be awarded a large sum of cash if you win. Betting on an even number would not be as rewarding since you would have a 50/50 chance. If the casino is fair (and none of them are, given how Thorp was kicked out of Las Vegas), then if you put down a dollar on snake eyes and you won, you should receive $36 back in return. Quite a nice return, you will agree!

Of course, you do not have to bet in a casino. You could invest your money into companies. The smaller, less known, and less likely a company is to succeed (given prevailing market conditions and business strategy), the more money you will get in return if it does succeed (for a fixed investment). This 'smells' very much like what we said of information: the smaller the probability of an event, the larger its information gain and in this case the larger the information the larger the payoff should be. Shannon's information principles are therefore, indeed, also at work in the business world. And betting on the success of companies is fairer than any casino (assuming efficient markets of course).

We all like to win. No one likes to be a loser. So can information theory guide us and maximize our chances of success? Astonishing though it may seem, the answer is a definitive 'yes'. Maximizing profit in financial speculations is exactly the same problem as maximizing the channel capacity for communication. The latter we know was already solved by Shannon in Chapter 3. So this is just a case of applying the solution to other problems. The solution was that the maximum capacity of a channel is achieved if the lengths of messages follow the 'log of the inverse probability' rule of Shannon's. If you now think of the amount of money that you are receiving on successful investments as the length of your message, than you obtain something that the (remaining) City of London and Wall Street traders refer to as 'the log optimal portfolio'.

Here is how this betting with the log optimal portfolio works in practice. The basic logic is to spread your bets as best as possible, i.e. don't place all your eggs in the same basket. The reason for this is that you would like to keep betting (and hopefully winning) for a long time. If all your bets are in one place, and you lose at the beginning (you might just be very unlucky) then you no longer have any credit to make the next bet. Just as in communicating, some of the bets will involve a large sum of money; some, on the other hand, will need to be very small. But how exactly should we determine the size of our investment?

Let's imagine that you like to play it safe and play for a long time – like a real connoisseur of gambling. You'd like to invest for 20 years, and whilst you are not looking for any huge return per year, you do want your money to add up nicely in the long term (so that you can enjoy your older age, as with pension funds). Here is what information theory tells us to do (in reality you will, most likely, get a financial advisor who will do all the thinking, calculating, and investing for you, but I would like to show you what kind of calculations they do. And remember, they also take a percentage – if you do it yourself you could be better off in the final analysis).

Let's take the top-10 FTSE (this is the London Financial Times Stock Exchange) companies at present. They are more likely to remain in the top-100 for quite some time and not to go under any time soon. Say you have a lump sum of $1000 that you would want to invest across these 10 companies. You first want to estimate the probabilities that these companies will make a positive return to you (i.e. you will win by investing in them). To estimate these probabilities (no one gives them to you) you can look at the past performance of the companies as well as current market conditions and strategies. The better your understanding of all the public information available about this company the more accurately you'll be able to estimate your probabilities and hence hope to derive an optimal betting strategy. This is the full-time job of City analysts. This is what they do 80 hours a week, every week (currently it's gone up to 90 hours a week, as a friend of mine reliably tells me). However even with only the basic level of research and approximate probabilities you can still make a good return by spreading your investment according to Shannon.

The share value of a company will sometimes go up and sometimes come down; in other words, there will be some natural fluctuation in the market. Overall, however, if the company is financially and strategically sound, and top-100 FTSE companies tend to be more so, then they will demonstrate an overall upward trend in the long term.

Using all this publically available information, we can estimate the probability that a company will make 3% next year. This is not as difficult as it initially looks (if we just want rough estimates). If it has made more than 3% every year for the past 50 years and analysts expect it to make 10% next year then the probability of it making 3% is fairly high. This way you can estimate the probabilities of certain returns for Company X.

Note that Thorp did not have the problem of estimating probabilities in playing blackjack, as the set of outcomes is well known, and by reading the cards he was able to recalculate probabilities. In fact, this is the exact reason why Thorp chose blackjack over other games such as slot machines (with a slot machine there is no prior information you could use to predict the next outcome). While playing blackjack Thorp was able to estimate future probabilities by seeing what cards had already been presented and adjusting his betting strategy appropriately. Following the performances of companies is like following cards.

In the highly unlikely event of the reader not being familiar with the game of blackjack (in which case I urge you to get out more) I'll give a brief overview. The basic premise of the game is that you want to have a hand value that is closer to 21 points than that of the dealer (without going over 21; if you do, you lose). Each card is worth a certain number of points and the dealer hands them out upon request. Other players at the table are of no concern, other than the fact that you can memorize their cards and adjust your probabilities appropriately. Your hand is strictly played out against the hand of the dealer. To determine whether to ask for another card, Thorp was simply using Shannon's formula into which he inserted the estimated probabilities.

Let's say we are playing against the dealer in a Vegas casino. We have three cards in front of us totaling 15 points and the dealer is asking us whether we want another card. So now we are thinking, what are the chances of the next card being six or less? Intuitively, given that we have already seen four sixes, three fives, three fours, and three twos, and three aces, we already have more information than somebody who wasn't paying any attention. Whereas somebody else may take the risk we know that logically there is a lower chance of seeing a card that will not take us over 21. So Shannon's formula just formalizes this and indicates when it is and isn't a good time to take a risk.

Walking into a casino may be a risky proposition altogether, especially if you are borrowing mafia money to gamble. So, let's think of a more wholesome scenario. After all why worry about casinos when gambling is fully legalized in the great global financial market of ours (I apologise to the reader for the slight cynicism as I sit here in October 2008 watching my pension fund taking a hammering).

We can extrapolate Shannon's formula and use it to invest in the stock market. Let us assume that one company has a 55% chance of a 3% return over a year, another has a 60% chance, and so on. How should we now invest? The best investment intuitively would be to invest proportionally. Namely you should invest most in the company that has highest returning odds, less into the next one, and so on. If one company has a twice higher chance of yielding the same positive return on your capital, then you should invest twice as much into it. And how about the return on your investment – how much would that be? The expected return would amazingly enough be exactly equal to Shannon's information formula! How can this be understood?

I would like to take a concrete example, just to show you how much you can objectively really earn in practice. Suppose that you are only investing into a top company and that it has a 55% chance of winning you money in a day. Say, as before, that you have $1000 to invest. First of all, information theory says that you should not invest more than 2 times 55 minus 100% ($2 \times 55 - 100$) of your money. This comes out to 10%, which is $100. The reason for this is that you need a buffer if you lose money – you would still like to be able to keep playing. It is only if you keep playing that the loot can add up to a substantial amount.

You have now invested $100. Shannon's information theory says that the rate at which your profit will increase is 0.5%. This means that you will have gained 50 cents. Next day, if things have not changed, you will also get 0.5%, but now you have started with $100 and 50 cents – which is more than the first time, being 50 cents and 0.5% of 50 cents, and so on…At the beginning you will not notice much change, but the growth is basically exponential (all exponentials are linear to start with which is why they do not look impressive initially). After one year – 365 days – you will end up doubling your money to $200. Next year you would have $400, then eight, and after five years you would have gained 16 times more than you started with. This rate of increase seems much better than the rate of inflation which is eating away at your capital if you are lazy with it and keep it safe in your bank and uninvested.

The earnings in the previous paragraph still may not sound all that much, but remember that you only invested a $100 in the first place. What if you'd invested $100,000? You would still earn 0.5% on top of it, but this would now be $500 (per day!). And all this for doing nothing, 'just (in this case very slightly) risking' the possibility of losing your money.

Shannon's formula will also tell us that you should not bet on anything that is completely uncertain (so don't play the lottery if you listen to Kelly's advice). If the chances of success are only 50%, then you would be wise to abstain from betting. On average you will gain nothing and in all probability you will lose everything.

Now let's link the discussion of financial markets and monetary speculations to the previous chapter, namely to the concept of Maxwell's demon. Can we use thermodynamics to describe the overall behaviour of markets and to deduce some very general trends in the financial world?

There is an obvious way in which thermodynamics affects markets. The price of something is clearly driven by the supply and demand principle and if the resources are scarce, then the price will be pushed up. As we have seen, thermodynamics implies that our energy resources will be getting depleted as well as causing other unpleasant effects on our planet as they wind down. This is bound to have an effect on the economy in general and we, in fact, see this kind of stuff around us all the time. The price of petrol is linked to oil and this in turn depends on the reserves of oil as well as on its availability. Even the slightest sniff of a conflict in the Middle East pushes up the world petrol prices.

However, thermodynamics and economics are more closely linked than this. The behaviour of financial markets itself, in some sense, quite closely mirrors the behaviour of physical systems in thermodynamics. There is a general law in finance that you hear repeated very frequently: 'that in an efficient market there is no financial gain without risk'. Anything worth doing must, according to this law, have a (significant) probability of failure associated with it. If something is a sure thing, you can bet that the reward is going to be negligible.

This sounds very much like the statement that in order to produce some useful work, you must be prepared to waste some heat. The Second Law of thermodynamics tells us that you must increase the temperature of your environment in order to do work and this is synonymous with the fact that the house takes more money than is proportionally fair (i.e. on average it always costs you more to win than you earn). Let us go back to the simple game of betting in a casino to illustrate this point further. We would like to argue that money will play the role of energy.

But, wait a minute – you will say, how can that be when money, unlike energy itself, is not quite conserved? Money can be generated out of nowhere while energy, as we have seen, cannot. If we have more people in the world, there will be more money around, so money is obviously not a constant. However, if we assume that we have a person betting against the casino in a house, then the total amount of money is fixed. The money could be in a person's pocket or it could be in the casino's safe, but it has to be somewhere. So, only under these specific conditions, is money conserved and can be compared with energy.

Now we can state the rules of betting in the casino in a way that we recover analogues of the laws of thermodynamics. First of all, the odds given by the house are such that:

1   'You cannot win the game'. This means that if the probability for something is one third, your return will at most be three times

your investment. So, if you invest $10 you will get at most $30 back, but with an overwhelming probability – two-thirds (67%) – you will lose. On average as you continue the game, you will not win anything. This is the exact analogue of the First Law of thermodynamics – you cannot get something for nothing.

The Second Law of betting is even more depressing, since casinos are designed in such a way that

2 *'You cannot even break even'.* This means that you will never get the fair odds, plus in a real casino you will have to pay to enter, you will have to pay to play, you will – most likely – be buying some drinks and tipping the casino personnel as you go along. All this contributes to the fact that while money is overall conserved it has the tendency to move from your pocket into the casino safe and not the other way round. This is the analogue of entropy increase in thermodynamics, where heat always flows from a hot to a cold body and never the other way round. So, in casino betting, the player is 'hot', while the casino is 'cool', and this determines the direction of the flow of money.

Now, there is also a Third Law of thermodynamics, which, of course, also has its analogue in betting. It says that:

3 *'You cannot quit the game'.*

I will leave the reader to interpret the meaning of this statement themselves. Suffice it to say that gamblers usually gamble till the bitter end. The laws of betting are just so very similar to the laws of thermodynamics (the Third Law of thermodynamics being the one prohibiting us from reaching absolute zero; in gambling terms this means being cooler than anything else, in particular the casino itself).

All the discussion so far can simply be summarized by saying that Shannon's formula for the maximum channel capacity,

Boltzmann's formula for physical entropy, and Kelly's formula for maximizing profit are all one and the same formula.

However, what is really surprising is that betting can be seen as a very fruitful metaphor for the evolution of life itself. We can think of random mutations as producing different gambling strategies, each with a different likelihood of success. This is exactly what a biologist, Frederic Warburton, analysed in one of his papers.

Warburton views evolution as a game between an individual (the gambler) and its environment (the house). Profiting in evolutionary terms means that, regardless of the environment, life will propagate. Losing means that life will end. So the game is as follows: the individual produces copies of itself, but naturally the copy will be slightly different due to random mutations. They may arise from the environment (due to cosmic rays, say) or some error in transcription. These mutations lead to a slightly altered set of properties in the new individuals, which are then tested by Nature and the environment. The number of individuals produced depends very much on the gambling strategy being employed. Any individual produced will have to be fed and nurtured and, given that resources are finite, we must choose how many copies to make in line with these restrictions. The new individuals then either pass the evolution test and further multiply (i.e. they generate profit) or they fail and die (i.e. the investment is lost). In economics, this is known as the principle of increasing returns. It is now clear that those that survive profit more and more in the sense that their strategy mirrors the environment better and better. Thus life increases profitability and evolution becomes a gambling game.

However, as always there is a snag; the more profitable life becomes the less profitable its environment (as it increases proportionally in entropy). As the environment increases in entropy, this makes it more and more difficult for life to propagate. In a similar manner, as you become more profitable in a casino (analogous to the environment) it in turn necessarily affects your profit in the long term through several consequences: it may reduce its payouts, charge a percentage tax on your winnings, or perhaps pack up altogether. Given that there is even a charge for playing the game in the first place, we can see how the second principle of betting manifests itself: namely that you cannot even break even.

We said that we can view life as a Maxwell's demon trying to keep low entropy locally, while increasing the entropy of its environment. In fact, as we saw, the more we keep the local entropy low the more we will keep increasing the total entropy of the system. Can the total entropy increase therefore be an indicator of life? Could we scan other planets for their entropy content to determine whether any life exists on them?

This is a fascinating question and the answer is that the total entropy production, while important, is probably not the only relevant factor. Out of all planets and moons in the Solar System, Mercury has the highest entropy production per unit area, but as far as we know, it has no life. The answer most likely lies in the fact that Mercury has no atmosphere and an atmosphere provides a medium for transport of matter that is crucial for living processes to take place.

The Earth and the Moon share the second place for the highest entropy production, which is roughly a quarter that of Mercury. The Moon again has virtually no atmosphere (it is gravitationally too weak to attract atoms to its surface) and this makes it difficult to have life. The rest of the planets and moons have entropy production far lower (Mars is fourth with two-thirds of the entropy production of the Earth or Moon).

Therefore, the total entropy production is very relevant for life, but it seems to be only a part of the story. However, there are other more dramatic options. It could also be that we have to enlarge our definition of life, and then we will recognize it in other places. Given, however, that it is very difficult to specify all the features of life (are viruses living or not?), entropy production seems to be quite a good single indicator.

Another important feature of life is that it becomes more complex. Let us see if we are able to quantify this trend more precisely, based on the discussion so far. Let us define complexity as the difference between a maximally disordered state, i.e. a maximum entropy, and the actual entropy with respect to the local environment (a thermo-dynamicist called Peter Landsberg actually proposed this defini-tion). Natural selection favours organisms that minimize entropy with respect to environment (the best gamblers in the language of this chapter). Those that don't, die out. Therefore, mutations will eventually lead to lower local entropy which, given that we subtract from the maximal total entropy to get complexity, results in increased complexity of life. It might be worth noting that it isn't that all life forms become more complex with time – obviously they don't and many are highly successful as they are and have been around for ages (e.g. types of bacteria). But more complex forms appear over time and it is this increase in some forms we are trying to explain.

The increase of complexity of life with time is now seen to be a direct consequence of evolution: random mutations and natural selection. In fact, the famous Oxford biologist Richard Dawkins has argued passionately in a number of his popular accounts that evolu-tion is the only theory we have that can explain all the living complexity we see around us. Given that we stated that life may maximize total entropy production this may imply that it also maxi-mizes complexity increase. Biological complexity is therefore the same as the successful gambler's profit and it grows in the same way.

Through analysis of betting and financial speculation and then synthesis of its fundamental principles, we can see another example of how information seems to be the most natural framework within which to discuss such ideas. We extend this approach again in the next chapter to discuss the social interactions that define how we live and the quality of our lives.

Key points

- It is possible to use Shannon's logic to drive betting strategies for maximizing profit.
- If you want to win in a casino, you must invest according to the probabilities for various gambling outcomes. Low probability should deserve a correspondingly low investment.
- Gambling on a completely random event is really gambling of the lowest order, and in the long run you'll never make a profit (given that half the time you win and half the time you will lose).
- If you follow Shannon's logic, your return on an investment will be according to Shannon's entropy. In finance, this way of spreading your bets is known as the log optimal portfolio approach and is widely used as the basis for investment decisions.
- Life is also a form of gambling, where life wins by propagating itself. The degree to which it wins is related to the length of time for which it propagates.

Social Informatics: Get Connected or Die Tryin'

Everybody knows a Joe. Joe is the kind of guy who was the most popular boy in class, head boy at school, the life and soul of the party, and whenever he needs something, it just seems to happen for him. This is the guy we love to hate! Why is he getting all the breaks when we have to work so damn hard? As we continue to grind out each day at work, we see Joe is the guy with a big house, fast car, and the most beautiful women swooning over him. Most men would give their right arm to have a bit of that magic.

So, how does he do it? Of course, I cannot tell you for sure (if I could my next book would be a bestselling self-help book), but it should come as no surprise that people with more friends and contacts tend to be more successful than people with fewer. Intuitively, we know that these people, by virtue of their wide range of contacts, seem to have more support and opportunity to make the choices they want.

Likewise, again it's no surprise that more interconnected societies tend to be able to cope better with challenging events than ones where people are segregated or isolated. Initially it seems unlikely that this connectedness has anything to do with Shannon's information theory; after all what does sending a message down a telephone line have to do with how societies function or respond to events?

The first substantial clue that information may play some role in sociology came in 1971 from the American economist and Nobel Laureate, Thomas Schelling. Up until his time sociology was a highly qualitative subject (and still predominantly is); however he showed how certain social paradigms could be approached in the same rigorous quantitative manner as other processes where exchange of information is the key driver.

Schelling is an interesting character. He served with the Marshall Plan (the plan to help Europe recover after World War II), the White House, and the Executive Office of the President from 1948 to 1953, as well as holding a string of positions at illustrious academic institutions, including Yale and Harvard. Schelling is most famous for his work on conflicts between nation-states, particularly those with nuclear weapons. Here the central idea is that of 'pre-commitment'. One party in a conflict can strengthen its strategic position by cutting off some of its options to make its threats more credible. For example, an army that burns its bridges, making retreat impossible, is a classic military example – here the opposition now knows you can no longer retreat and therefore its strategy is more limited. They know you are fighting until the end – a classic case of brinkmanship.

But Schelling went further. He realized that studying conflict can result in an understanding stretching far beyond military conflict. He applied a similar analysis to individuals' internal struggles. The problem, he suggested, is that pretty much everybody suffers from a split personality on various issues. For example, one side of you may desperately want to lose weight or quit smoking or run two miles a day or get up early to work. On the other hand, the other side may want an extra dessert or a cigarette, hate exercise, or love sleep. Both selves are equally valid, and equally rational about pursuing their desires. But they do not exist at the same time and which side wins depends on the strategies that the two personalities use. In Schelling's view, we could improve the chances of one side winning by showing a huge commitment which the other side would find difficult to match. So whilst the second side says I want to stay in bed a bit longer, the first side can retaliate with 'but I am paying $70 a week for morning personal training sessions'. Gym memberships, not keeping cigarettes in the house, not having a car and walking to work instead, are all examples of what Shelling referred to as 'burning your bridges'.

Studying conflict thus turned out to be important not only for the military, but also for individuals and, as we will see shortly, for understanding segregation in a society. And underlying it will, of course, be the concept of information.

The idea of using information theory in social studies is certainly not something new and we can trace its roots back way before Schelling. In fact, it is the use of statistical methods in social sciences that prompted Boltzmann to apply them within physics in order to come up with his entropy formula. Of course, a direct application of the statistical methods used in information theory is much more challenging given the complex nature of any human society (yes, human society is far more complex than any physical system), but it has the same underlying premise. When compared to a simple gas of atoms, which is a typical model in physics, every human being has added variability in that he or she can formulate an independent complex strategy in order to satisfy their own ends.

Information itself appears in many different guises in the social context: connections between individuals, actions, and states of individuals, capacity of societies to process data, and so on. All of these types of information play a role in the functioning of a society.

An important idea that we have alluded to in the previous chapters, but whose more precise introduction we have been delaying, is that of 'mutual information'. This concept is very important in understanding a diverse number of phenomena in Nature and will be the key when we explain the origin of structure in any society.

Mutual information is the formal word used to describe the situation when two (or more) events share information about one another. Having mutual information between events means that they are no longer independent; one event has something to tell you about the other. For example, when someone asks if you'd like a drink in a bar, how many times have you replied 'I'll have one if you have one'? This statement means that you are immediately correlating your actions with the actions of the person offering you a drink. If they have a drink, so will you; if they don't, neither will you. Your choice to drink-or-not-to-drink is completely tied to theirs and hence, in information theory parlance, you both have maximum mutual information.

A little more formally, the whole presence of mutual information can be phrased as an inference indicator. Two things have mutual information if by looking at just one of them you can infer something about one of the properties of the other one. So, in the above example, if I see that you have a drink in front of you that means logically that the person offering you a drink also has a glass in front of them (given that you only drink when the person next to you drinks).

Of course mutual information between events is not always perfect, it could also be imperfect, in that you 'usually' but not always will have a drink if your friend has a drink. In this case any inference between the two of you will be weaker.

Whenever we discuss mutual information we are really asking how much information an object/person/idea has about another object/person/idea. In the telephone communication example, Alice and Bob share mutual information. After she has communicated to Bob, the two now share this same message. When Bob tells you what the message is, assuming maximum mutual information between Alice and Bob, you will then know what message Alice sent without having to ask Alice. In cases of non-maximal mutual information (e.g. Bob may have forgotten parts of it) we can only infer the message with partial success.

When it comes to DNA, its molecules share information about the protein they encode. Different strands of DNA share information about each other as well (we know that A only binds to G and C only binds to T). Furthermore the DNA molecules of different people also share information about one another (a father and a son, for example, share half of their DNA genetic material) and the DNA is itself sharing information with the environment – in that the environment determines through natural selection how the DNA evolves.

In thermodynamics, mutual information is established between the demon's memory and the movement of the particles. How much work the demon can extract is dependent on how much information the demon has on the actual speed of the particles. In financial strategies, the profit derived is inevitably linked to how much information you share with the market (random betting aside). If you want to know how the price of the stock will move, then you need to have complete information about the market (often virtually impossible given the complexity and volume of information). Given that everything interacts to some degree with everything else, mutual information is simply everywhere!

One of the phenomena we will try to understand here, using mutual information, is what we call 'globalization', or the increasing interconnectedness of disparate societies. Another social phenomenon we will address is the division of every society into different classes and the related negative effects of segregation.

Before we delve further into social phenomena, I need to explain an important concept in physics called a phase transition. Stated somewhat loosely, phase transitions occur in a system when the information shared between the individual constituents become large (so for a gas in a box, for an iron rod in a magnetic field, and for a copper wire connected into an electric circuit, all their constituents share some degree of mutual information).

A high degree of mutual information often leads to a fundamentally different behaviour, although the individual constituents are still the same. To elaborate this point, the individual constituents are not affected on an individual basis, but as a group they exhibit entirely different behaviour. The key is how the individual constituents relate to one another and create a group dynamic. This is captured by the phrase 'more is different', by the physicist Philip Anderson, who contributed a great deal to the subject, culminating in his Nobel Prize in 1977.

A common example of a group dynamic is the effect we observe when boiling or freezing water (i.e. conversion of a liquid to a gas or conversion of a liquid to a solid). These extreme and visible changes of structures and behaviour are known as phase transitions. When water freezes, the phase transition occurs as the water molecules becomes more tightly correlated and these correlations manifest themselves in stronger molecular bonds and a more solid structure.

The formation of societies and significant changes in every society – such as a revolution or a civil war or the attainment of democracy – can, in fact, be better understood using the language of phase transitions.

I now present one particular example that will explain phase transitions in more detail. This example will then act as our model to explain various social phenomena that we will tackle later in the chapter. Let us imagine a simple solid, made up of a myriad of atoms (billions and billions of them). Atoms usually interact with each other, although these interactions hardly ever stretch beyond their nearest neighbours. So, atoms next to each other will feel each other's presence only, while the ones that are further apart from each other will typically never directly exchange any information.

It would now be expected that as a result of the 'nearest neighbour' interaction, only the atoms next to each other share information while this is not possible where there is no interaction. Though this may sound logical, it is in fact entirely incorrect. Think of a whip: you shake one end and this directly influences the speed and range at which the other end moves. You are transferring movement using the interconnectedness of atoms in the whip. Information can be shared between distant atoms because one atom interacts with its neighbours, but the neighbours also interact with their neighbours, and so on. This concept can be explained more elegantly through the concept of 'six degrees of separation'.

You often see it claimed that each person on this planet is at most six people away from any other person. This sounds shocking at first but actually makes complete sense if you look at it in detail. I am, for example, just three degrees of separation (three people) away from Bill Clinton. How's this? Well, I know a guy quite well who works in the City of London. His boss knows the boss of the American branch of the bank and the American boss knows Bill Clinton. So there we go! So if you count the number of people separating Bill Clinton and me you will see that there are only three. Though I may be particularly indifferent to the fact that I am separated from former President Clinton only by three degrees (he surely is), it is nevertheless interesting that this is so. It is also interesting that this is how connections work. It doesn't matter whether you know a person doing X, all that matters is that you are connected into society, because, through this, you have access to almost everybody on the planet (and somebody somewhere should know something about X).

It is very easy to see why any one person is connected to any other with at most six degrees of separation. Say that I know roughly 100 people (probably more, but it is said that 200 is the capacity for the human brain to differentiate and memorize different names and faces – this is our evolutionary inheritance for the early days of human evolution, where time was spent in small tribes). Each of the 100 people I know knows 100 more people, and so on. After five steps this number reaches ten billion (100 times 100 times 100 times 100 times 100, five times). The number of people in the world is six billion. So, even five steps of connections and acquaintances are enough in this model to connect everyone to everyone else on this planet. But, of course, maybe not everyone knows 100 people, and so on, which leads to perhaps needing six degrees to make a network that covers everyone in the world. Perhaps this is why in society we gravitate to more connected people, to optimize our interconnectedness.

Why is this networking between people important? You might argue that decisions made by society are to a high degree controlled by individuals – who ultimately think for themselves. It is clear, however, that this thinking is based on the common information shared between individuals. It is this interaction between individuals that is responsible for the different structures within society as well as society itself.

Societies can, in addition, have a certain number of other features, such as a degree of centralization in their decision making. One extreme form of centralization is dictatorship, and it is not very interesting to analyse as individuals have less of a role in any decision making (by definition). The other extreme is much more interesting, and this means that there is no centralization and where individuals line up of their own accord. Here we would like the consensus to emerge just by individuals interacting with one another – no external stimulus is allowed. In this case, the information shared between individuals becomes much more important. So how do all people agree to make a decision, if they only interact locally, i.e. with a very limited number of neighbours?

In order to understand how local correlations can lead to the establishment of structures within society, let us return to the example of a solid. Solids are regular arrays of atoms. This time, however, rather than talking about how water becomes ice, let's consider how a solid becomes a magnet. Every atom in a solid can be thought of as a little magnet on its own. Initially these magnets are completely independent of one another and there is no common north/south alignment – meaning that they are all pointing in random directions. The whole solid – the whole collection of atoms – would then be a random collection of magnets and would not be magnetized as a whole (this is known as a paramagnet). All the random little atomic magnets would simply cancel each other out in effect and there would be no net magnetic field.

However, if the atoms interact, then they can affect each other's state, i.e. they can cause their neighbours to line up with them. Now through the same principle as six degrees of separation, each atom affects the other atoms it is connected to, and in turn these affect their own neighbours, eventually correlating all the atoms in the solid. If the interaction is stronger than the noise due to the external temperature, then all magnets will eventually align in the same direction and the solid as a whole generates a net magnetic field and hence becomes magnetic! All atoms now behave coherently in tune, just like one big magnet. The point at which all atoms 'spontane-ously' align is known as the point of phase transition, i.e. the point at which a solid becomes a magnet.

Can this simple idea be applied to something as complicated as human society? Initially, this seems like it may be very difficult. We would like to think of humans as little magnets (let's say that they can 'politically' point in the left or right direction) and we now inves-tigate under what circumstances whole societies can swing to the left or to the right and how this requires very little external influence. This would be the social analogue of the magnetic phase transition.

You may object that atoms are simple systems compared to humans. After all humans can think, feel, get angry, while atoms are not alive and their range of behaviour is far simpler. But this is not the point! The point is that we are only focusing on one relevant property of humans (or atoms) here. Atoms are not all that simple either, but we are choosing to make them so by looking only at their magnetic properties. Humans are much more complicated still, but now we only want to know about their political preference, and these can be quite simple in practice.

Let us discuss this aspect a bit more. Surely there is a crucial difference between atoms and humans. Namely, humans have free will. They are autonomous and make their own decisions, create their own destinies. Atoms are incapable of this. While this discussion of free will is a very intricate one, and we will debate it later, there is something that could already be said on it here.

Suppose for that matter that you observe satellite pictures of people walking up and down a busy street, such as Oxford Street in London. You will not be able to make out who the individuals are – that is if you use the google.com satellite pictures (I am sure that CIA or MOSSAD can easily accomplish a better resolution). People will appear as dots moving between two straight lines – the edges of the pavement of Oxford Street. There will be some larger dots in the middle representing the traffic, which we will ignore for the time being.

You now observe how the dots move. They will broadly move up and down the street, but there will be a certain movement to the left and to the right too, as the dots avoid each other. Imagine now that someone shows you this picture of dots moving in this fashion, but does not tell you that they represent people. Would you be able to tell, just from the motion of the dots, whether they represent something living (even intelligent?) from something non-living (such as atoms, or billiard balls hitting each other and bouncing off in different directions)?

Most likely you wouldn't be able to tell the difference, as the motion of people or atoms would be roughly the same (in their own ways they both avoid collisions, either via seeing and moving or via electrostatic repulsion). Both would be known as a diffusion process – a general overall flow in one direction (up and down), but with some random irregularities every once in a while. And this is precisely my point, that living and non-living things appear to behave in a very similar manner.

We now return to the issue of phase transitions and human societies. Suppose that we have a chain of systems where each system only interacts with its two neighbours. Such a system was first considered in physics by Ernest Ising, who studied it during his PhD in the 1920s. Unfortunately, he was able to prove very conclusively that there is no phase transition in such a model. So, if you think of atoms in such a chain as being little magnets, then they can never ever spontaneously align with each other no matter what happens outside. This was apparently such a disappointing conclusion to Ising, who was hoping to explain phase transitions microscopically, that he quit physics after his PhD!

Too bad for Ising, because some 20 years later Lars Onsager showed in a very beautiful paper that if instead of a chain you look at a two-dimensional array of atoms, then there is a phase transition at low enough temperatures. And for this, Onsager was awarded a Nobel Prize for chemistry. The whole area of phase transitions then started to thrive, resulting in a number of fundamental discoveries and Nobel Prizes (Anderson's included).

Among one of the most striking results in phase transitions that I am aware of is a very general 'no-go' theorem. No-go theorems are broad results that rule out some specific kind of behaviour. Ising's result is a special kind of no-go theorem that says that there are no phase transitions in one dimension (chains). Interestingly, it can be shown that there are no general phase transitions in two dimensions (planes) either. The only exception to this no-go in the two-dimensional case is the model that Onsager analysed (so he was very lucky after all!). The very fact that we see ice turn into water and water into vapour already implies that the Universe has to be at least three-dimensional.

Anyway, what has all this to do with human societies? You can think of atoms in the above systems, not as little magnets, but as human beings! Yes, as commented above, we are more complicated than atoms, but some of our features are very simple. Think of a human being and his political predispositions. Suppose that we have two choices in a society, to be conservative or to be liberal. These two states would be the human analogue of a magnet with two states: pointing to the right, or pointing to the left (I am fully aware that some cynics might say that 'in power' or 'out of power' are more important directions in politics).

Now, in general, a society will be in a state such that some members are conservative and some liberal. But suppose that we ask under what conditions would all members become, say, liberal? Our discussion of Ising's result would suggest that if we have a very simple society where every member discusses politics only with their two nearest neighbours (and no one else), then there is no way that we can have a fully liberal society. We can never have a spontaneous liberalization and we are always forced to live with some conservatives!

A one-dimensional society is therefore very easy to describe and there are no surprising changes here. One such society is the mafia depicted in the movie *Goodfellas* (this is indeed far from being liberal). The main boss of the mafia, the capo di tutti capi, is a guy called Pauly, who would only communicate his orders to his two closest men – he would rarely talk to anyone else. They would then in turn communicate to their closest allies, and so on. This is not only secure so far as it prevents the FBI from finding the key people in the mafia, but it also leads to an increased stability within the mafia itself. It is therefore very unlikely that the mafia as a whole will suddenly undergo a phase transition and change its behaviour significantly, for example, electing a new leader or breaking down due to a few rogue informants.

Interestingly, if we allow everyone to interact with everyone else, then even in one dimension phase transitions are possible. Now, the real world is somewhere between these two extremes. None of us really speaks only to our neighbours, and no-one certainly communicates to everyone in the world. As I said above, our circle of family, friends, and acquaintances to which we are sufficiently connected usually extends to roughly 200 people.

In fact, the number of connections we have with other humans follows what is known as a power law distribution. The number of people who know lots of other people is smaller than the number of people who know few. And the ratio of the two numbers follows a strict law, similar to the Zipf law of letter frequencies mentioned earlier in the book. More precisely, there are a million times fewer people with 1000 contacts than with 10 contacts. People with lots of connections are just statistically improbable in the same way that longer words are less likely to be heard in English.

This unevenness in the number of contacts leads to a very important model where there is a great deal of interaction with people close by and then, every once in a while, there is a long-distance interaction with someone far away. This is called a 'small world network' and is an excellent model for how and why disease propagates rapidly in our world. When we get ill, disease usually spreads quickly to our closest neighbours. Then it is enough that only one of the neighbours takes a long-distance flight and this can then make the virus spread in distant places. And this is why we are very worried about swine flu and all sorts of other potential viruses that can kill humans.

Let us now consider why some people believe – rightly or wrongly – that the information revolution has and will transform our society more than any other revolution in the past – such as the industrial revolution discussed in earlier chapters. Some sociologists, such as Manuel Castells, believe that the Internet will inflict much more profound transformations in our society than ever previously seen in history. His logic is based on the above idea of phase transitions, though, being a sociologist, he may not be interpreting them in quite the same way as a physicist does mathematically.

To explain, we can think of early societies as very 'local' in nature. One tribe exists here, another over there, but with very little communication between them. Even towards the end of the nineteenth century, transfer of ideas and communication in general were still very slow. So for a long time humans have lived in societies where communication was very short range. And, in physics, this would mean that abrupt changes are impossible. Societies have other complexities, so I would say that 'fundamental change is unlikely' rather than 'impossible'. Very recently, through the increasing availability of technology we can travel far and wide, and through the Internet we can learn from and communicate with virtually anyone in the world.

Early societies were like the Ising model, while later ones are more like the small world networks. Increasingly, however, we are approaching the stage where everyone can and does interact with everyone else. And this is exactly when phase transitions become increasingly more likely. Money (and even labour) can travel from one end of the globe to another in a matter of seconds or even faster. This, of course, has an effect on all elements of our society.

Analysing social structures in terms of information theory can frequently reveal very counterintuitive features. This is why it is important to be familiar with a language of information theory, because without a formalized framework, some of the most startling and beautiful effects are much harder to understand in terms of root causes. Take segregation in a modern society, for example. Modern-day Los Angeles has distinct areas with clearly defined boundaries for white, Hispanic and black Americans. It is clear that this segregation is not legally imposed from above (each person is legally free to buy a property in any area they wish to) so is the inescapable conclusion that they are all a bunch of racists? The surprising answer, according to Schelling, is no, and that even the most liberal minded communities could end up transitioning into a segregated state. Segregation in this sense occurs very much like the phase transition we discussed earlier.

To simulate segregation, Schelling used a very simple model analogous to the phase transition model of a magnet described before. Imagine a two-dimensional grid, on which you can lay black and white pieces randomly. If we take this as synonymous with communities, this corresponds to a very liberal and mixed community (in the magnet example, this corresponds to a completely disordered state, where all the little magnets are aligned in all different directions).

Now we have to study the dynamics and see how segregation emerges. Schelling looked at the following rule: each and every piece looks around at its neighbours and if it sees a certain number of neighbours of different colour, it makes a decision to move somewhere else. It's clear that an extreme racist attitude is going to lead to segregation (e.g. if every piece moved just because one of its neighbours were of a different colour). However, and this is the surprise Schelling presented, it appears that even if you take a very liberal view, and move only when all your neighbours are of a different colour, this still leads us naturally to a segregated society. What this means in the Ising model is that even the weakest interaction is enough to make it undergo a phase transition. In such a model, interactions would force two spins to align themselves in the same way that in Schelling's model neighbours become racially homogeneous (to remind the reader, a city is analogous to a two-dimensional structure and in two dimensions, unlike in one dimension, the Ising model exhibits phase transitions).

The most natural way of thinking about this is in terms of mutual information. In the initial state, which was completely disordered, there was very little mutual information – because by looking at one piece, it is difficult to infer the colour of the neighbours (as by definition, the pieces are laid out in a disordered manner). However the mutual information is maximal in a maximally segregated society, as you can look at one piece and you can then infer the colour of all the pieces surrounding it. As with other phase transitions we have talked about, here mutual information is also a signature of a segregated society (a segregated society would represent an ordered state of magnets, when they all behave collectively).

If you properly understand Schelling's simple model, then you will quickly see that this doesn't have to apply to just racial segregation. It could also apply to any other grouping in a society. Political, financial, social, intellectual separations can all be studied and understood using the same logic.

We have been grossly simplifying societies here, but our analysis certainly contains some elements of truth. To see how much truth, let us test our predictive power on a very simple question. What is the distribution of wealth in a typical society? This question combines the gambling elements we discussed in the previous chapter with social elements we are discussing now.

If the basis of every society is information, as I have been arguing, does optimizing information tell us about the wealth distribution within a society? The answer is both yes and no. Let me explain the no first.

When we optimize the Shannon information we typically obtain a distribution of probabilities that is called Gaussian or bell curve. This distribution is named after the German mathematician Karl Friedrich Gauss, who first noticed the ubiquity of Gaussian distributions in Nature. For example, atoms in a gas distribute their speeds according to a Gaussian distribution. Most atoms move at some middle-range speed (say 500 metres per second). Then equally smaller numbers move at around 400 and 600 meters per second. The number of atoms that we record as we move away from 500 metres per second diminishes very quickly. And this is the main feature of Gaussians, which have a typical bell-shaped profile.

The distribution of wealth in most societies is not Gaussian. It is in fact a power law. This means that the number of very wealthy individuals is very small, whereas the number of poor is huge. If wealth followed a Gaussian distribution, then most people would be somewhere in the middle with small deviations to the richer and poorer side of the spectrum. This, alas, is never ever the case in any society we are aware of (not even in totalitarian communist regimes of the past such as the Soviet Union).

Where does the power law distribution come from? Why is the probability of having one million dollars in your bank account 1000 times smaller than having $1000? This is exactly what the power law says. If there are 100 millionaires in a country, then there are 100,000 people with only $1000.

Let us try to see why this is so. The complete answer is not known, but we can make some guesses (if I knew the answer I would probably immediately book myself a plane ticket to Stockholm to pick up a Noble Prize in economics). The favoured explanation seems to be a principle that can be called 'the-rich-get-richer'. In simple terms, wealth doesn't just add to wealth, it multiplies. Those that have more will get proportionally more and so the gap between the haves and have-nots increases to conform to the power law. Even if everyone starts equally well off, the small random differences in people's fortunes (some will win on the random stock market, others will lose) will sooner or later grow to become large.

Does this mean that information has nothing to do with wealth distribution? No. People acquire wealth not only by themselves, but through a set of networks that they create using all sorts of social mechanisms. Social interactions are complicated and go on in parallel in all corners of any given society. The social dynamic is very intricate and changes on very short time-scales.

This means that we expect information not to have the additive property, namely that the joint information contained in two independent events must be equal to the sum of the information content of each individual event (this was used in Chapter 3 to derive Shannon's entropy). If additivity fails, the correct measure of information is no longer Shannon's. One person who studied these non-additive measures extensively is a Brazilian physicist, Constantino Tsallis. Why is wealth not additive? As we have seen, having some wealth and gaining more does not just result in the sum of the two. Wealth does not just simply add, it super-adds! This is another view of the principle of increasing returns in economics (which we met in Chapter 6). There is no contradiction here, since wealth is not a conserved quantity. It can be created out of nothing (by nothing I mean 'no initial wealth'; I do not mean without any physical resources).

If we use a different formula to Shannon's, but still an inverse function of probability, we can explain the power law distribution. Therefore, even one of the key aspects of social dynamics and stability – its wealth distribution – is an outcome of simple informa-tion theory. If not Shannon's in actual form, this surely is Shannon in spirit.

Some sociologists are optimistic that the information age will lead to a fairer society that will improve everyone's living condi-tions, as well as narrowing the gap between the haves and have-nots. Others are rather pessimistic, claiming that the new age will bring an abrupt end, a kind of phase transition, to present society (through all sorts of mechanisms such as increased crime due to the break-down of families, global terrorism, global warming, and so on).

It is unlikely that information theory will tell us what the future has in store for us. One thing is for sure, though. With greater inter-connectedness of the world, we had better improve the speed of our thinking and decision making. In a more interconnected society we are more susceptible to sudden changes. Mutual information simply increases very rapidly and if we want to make good decisions we need to ensure that our own information processing keeps pace. The future, it seems, will not only favour the brave but also the fastest.

Having seen how information underpins various social, biolog-ical, and physical phenomena, we are now ready to take the discus-sion to the next level. We travel back to the questions of where this information comes from, how much of it there is in the Universe and what is the fastest that it can be processed. The answers to these questions underpin all the chapters we have seen so far. In order to understand the ultimate origins of information we need to take an exciting voyage of discovery. And this will take us into the realms of quantum mechanics, the true nature of randomness, whether teleportation is possible, and the question of free will and determinism. It's going to be a rocky boat, so hold on tight!

Key points

- Correlated systems are those that share some common information.
- Individuals in a society become correlated in many ways thorough various interactions, by communication for example.
- Societies are networks of correlated individuals. The degree of correlation and networking determines what kind of society we have.
- Most societies exist between the two extremes: completely connected and completely disconnected individuals. Such societies are an example of small world networks and explain many social phenomena, from the spread of disease to the winner-takes-all and fit-get-rich principles.

PART TWO

Up until now we have been discussing various aspects of reality in terms of information. We have seen how information theory provides a powerful description of various different aspects of reality. In deriving his theory, Shannon was just trying to describe information in terms of a very specific problem, that of communication between two parties, Alice and Bob. He didn't care a jot about the applicability of his theory beyond that problem, but as it turns out this is exactly where the power of his theory lies. Did Shannon get lucky or is there something more fundamental about his approach that makes information theory so widely applicable?

At its core, information theory asks the most fundamental question that can be asked: Is event A different (distinguishable) from event B? But why is this so fundamental, you may ask? Well, think of it this way; try to imagine learning or describing anything where you cannot distinguish between the correct answer and an incorrect answer. It cannot be done! You are essentially blind. Without distinguishability we cannot hope to have any understanding of our Universe if everything looks identical. If we consider the first half of this book we have already seen this concept of distinguishability applied in many different guises. In biology for example, DNA distinguishes between four bases to replicate itself. In thermodynamics, Maxwell's Demon needs to distinguish between fast-moving and slow-moving particles in order to create a temperature differences between two sides of a container. In a casino, you're betting on at least two different scenarios and it is your understanding of the probabilities that determines your profit. In sociology we were making a distinction between whether an individual is liberal or not, which has an impact on how likely it is that you stay in your neighborhood.

This fundamental concept of distinguishability between two different states is basically what Shannon referred to as a bit of information. A bit is the most fundamental measure of information; where you have more than two outcomes, you simply use more bits to distinguish them all.

Of course it makes no sense to talk about the distinguishability of two events if only one of the events ever occurs. Therefore we also need to know the probability with which each of the events occur. Because we can distinguish between more than one outcome, time after time, we can then generate the probabilities for these outcomes. The probability of an event gives us the expectation of an event occurring, and this is what allows us to quantify how surprised we are when the event happens. If we are expecting something with high probability we are not so surprised when it happens (e.g. the Sun rises) and when it doesn't happen, we are a little more surprised.

All of this is so natural and basic that even if you sat in your armchair and thought about it (albeit with a fair modicum of hindsight) you find yourself leaning towards a similar framework to that which Shannon proposed. From this perspective, I think you'd agree that Shannon's concept of information seems quite intuitive.

So are we done? Armed with Shannon's information theory can we now generalize any problem in terms of information? Well, not quite. While the key to wide applicability of Shannon's information theory is that it has the same logical foundation as our everyday physical, biological, social, and economical phenomena, it doesn't quite capture everything. Shannon's information theory boils down to events based on Boolean logic, i.e. for an event with several outcomes, each outcome happens or does not happen – if you roll a die you can either get 3 or not (you can't roll a 3 and a 6 at the same time). Note that though this logical foundation seems an entirely trivial truism to the reader, we will see that our latest understanding of the Universe tells us otherwise.

This event-based Boolean logic is also very characteristic of the early models of physics. In the same way that Shannon's information theory is based on the fundamental idea of a definite outcome (we only expect to see event A or event B occurring, but never the two together), our early physical models follow suit. The early models of physics, usually called classical or sometimes Newtonian physics, are also responsible for the vast majority of our technological progress to date (think: electromagnetism, energy manipulation, hydrodynamics, and telecommunications). It is the kind of physics that we studied at school, where each particle has a well-defined charge, position, velocity, and mass. Classical physics argues that if we knew these properties for all the particles in the Universe, we could determine the outcome of any future event, in other words the Universe is fully deterministic. In fact there has been an imaginary creature conceived for this very purpose. Laplace's demon, as he is affectionately referred to, knows all the properties for each and every particle in the Universe and as such is able to determine any event (akin to being able to see into the future). To this demon, the concept of information is completely redundant, he has no need for it – as the demon already knows with certainty what will occur. What a boring Universe this demon lives in. Samuel Johnson would agree: 'such is the state of life that none are happy but by the anticipation of change'.

Fortunately, though, even if this deterministic Universe is boring and predictable to Laplace's demon, scientific progress, and physics in particular, is still full of surprises. Physics is a very dynamic activity, and as soon as we are set on a model to describe reality, along comes an experiment that completely challenges our view. In this way, physics evolves through time, taking into account more and more information, new experiments and insights to produce a better and better description of reality. So the question is, what has made classical physics 'classical'? I mean who's the new kid on the block?

Our latest understanding is that certain aspects of reality are more accurately described by a higher approximation of physics known as 'quantum theory'. With quantum theory the notion of a deterministic Universe fails, events always occur with probabilities regardless of how much information you have. Hence Laplace's demon cannot even exist in principle, let alone in practice (i.e. there is no truth in a crystal ball or the Greek oracle at Delphi). It is not just physics that is full of surprises, the Universe intrinsically is full of surprises.

So what does our new understanding of physics, the introduction of quantum theory, mean for Shannon's information theory? Shannon didn't build his information theory to include the oddity of quantumness, so if we are really to describe the whole Universe, does the fundamentality of Shannon's information theory still hold? Yes, but with a twist. Distinguishability and probabilities are still central even in quantum theory, however the concept of distinguishability needs to be enlarged to account for the bizarreness of quantum effects. And, when we expand Shannon's information theory, we find that there are many additional elements of reality that had been completely hidden from us.

In the second part of this book we present and apply an enlarged information theory, quantum information theory, to present a number of novel features that couldn't be described purely by Shannon's information theory. Quantum information theory can be approximated by Shannon's information theory in certain cases, including the cases covered in the first part of the book. In the second part of the book we no longer have this luxury; a full quantum exposition is necessary. This leads us to some fascinating new features of information processing, such as the ability to teleport information across vast distances, the ability to compute faster than we ever thought imaginable, and to communicate so securely that no amount of eavesdropping or computing power can ever decipher our communications. However, the real surprise is how our understanding of the Universe is dramatically shaken and leads us to find new insight into its very origin.

Former US Secretary of State, Donald Rumsfeld, unbeknown to him, paraphrases the transformation from Shannon's to quantum information theory beautifully when he tells us that some of the unknown unknowns have now become known unknowns or even known knowns. This does not exclude the fact that there may be many more unknown unknowns, i.e. we still don't know what we don't know. Essentially, there's still much to look forward to!

Fortunately the logic of quantum mechanics is not quite as confusing as Rumsfeld's speech. In Part Two I show that it is possible to describe quantum information in a clear and concise manner to fit our latest description of reality.

Quantum Schmuntum: Lights, Camera, Action!

Spring 2005, whilst sitting at my desk in the physics department at Leeds University, marking yet more exam papers, I was interrupted by a phone call. Interruptions were not such a surprise at the time, a few weeks previously I had published an article on quantum theory in the popular science magazine, *New Scientist*, and had since been inundated with all sorts of calls from the public. Most callers were very enthusiastic, clearly demonstrating a healthy appetite for more information on this fascinating topic, albeit occasionally one or two either hadn't read the article, or perhaps had read into it a little too much. Comments ranging from 'Can quantum mechanics help prevent my hair loss?' to someone telling me that they had met their twin brother in a parallel Universe, were par for the course, and I was getting a couple of such questions each day. At Oxford we used to have a board for the most creative questions, especially the ones that clearly demonstrated the person had grasped some of the principles very well, but had then taken them to an extreme, and often, unbeknown to them, had violated several other physical laws on the way. Such questions served to remind us of the responsibility we had in communicating science – to make it clear and approachable but yet to be pragmatic. As a colleague of mine often said – sometimes working with a little physics can be more dangerous than working with none at all.

'Hello Professor Vedral, my name is Jon Spooner, I'm a theatre director and I am putting together a play on quantum theory', said the voice as I picked up the phone. 'I am weaving elements of quantum theory into the play and we want you as a consultant to verify whether we are interpreting it accurately'. Totally stunned for at least a good couple of seconds, I asked myself, 'This guy is doing what?' Had I misheard? A play on quantum theory? Anyway it occurred to me that there might be an appetite for something like this, given how successful the production of *Copenhagen*, a play by Michael Freyn, had been a few years back. *Copenhagen* was based on an actual meeting that took place in 1941 between two of the early fathers of quantum theory, the Danish Niels Bohr and the German Werner Heisenberg. In *Copenhagen* (the play) the efforts to clarify or communicate the subtleties of quantum theory were notable but this was not its purpose, so Spooner's play would provide a new spin on things. So I thought, why not? After exchanging a few details we agreed to meet the next morning to discuss it a little more.

One of the attractions of finding out more was that so often science and the arts are viewed as quite antagonistic to one another and it was interesting to find out how someone was trying to bridge that divide. It was also helpful that the building where they were rehearsing was only 200 yards away from my office (and who says theoretical physicists aren't practical!).

The following morning I dropped off my bags at the office and made my way to the theatre to see Jon. The theatre was located in one of the most historic sites on the east side of the university. On the same site is the Trinity St David's congregational church, built by G.F. Danby in 1898, which represents one of the rarest and prettiest examples of gothic architecture in central Leeds. It is a sterling indictment to the enterprise of the university that this church is now home to an impressive student theatre as well as one of the finest nightclubs in Leeds (Loiners will know it as *Halo*)! Some have commented that this seems to be a part of an ongoing trend, that actors and clubbers appear to be the main groups interested in organized religion these days.

On my first visit, I was a little nervous, not knowing much about theatre and what they expected from me. I was thinking that perhaps I could, in the worst case, just reel off an introductory lecture on quantum theory and brought along some printed diagrams just in case. As I entered the main hall I found several guys huddled in a corner, deep in active discussion. As they looked up, they saw me and I waved back. 'Ah Professor Vedral, come in and sit down. We have a few questions for you'. It seemed likely that my presence was related to whatever they were discussing, as the mere sight of me caused a decent amount of stirring and a whole-hearted diversion of attention. Dispensing with formalities they got straight down to business and asked me about the plausibility of a character in the play who was required to be in two places at the same time. They said that they had heard a little bit about quantum theory and teleportation and they wanted to know if this was possible.

While a couple of the crew seemed to have bought into this idea, several were still sceptical as to whether there was any basis in quantum theory for such a preposterous notion. However, all were understandably quite shocked when I revealed that this was part of my everyday work (though much to their disappointment I had to clarify that I was not experimenting with humans).

I went on to explain that at the heart of quantum physics is the concept of indeterminism. Indeterminism is linked to the fact that an object can indeed be in more than one state at any one time (e.g. you toss a coin and it can give you both heads and tails at the same time). In technical parlance this is known as a quantum superposition. The main difficulty with understanding and teaching quantum mechanics is exactly this counterintuitive and paradoxical nature – people just don't buy it first time around (even Einstein went to his grave certain that this was wrong).

In physics there is a very simple experiment demonstrating superposition and this is what I went on to describe to the theatre crew. Imagine a photon, which is a particle of light, encountering a beam-splitter. A beam-splitter is just a mirror that has a silver coating on it and, by varying the amount of silver, we can tune the probability with which the photon is reflected or transmitted. Let us say that these probabilities are made equal, so that we have the exact equivalent of a fair coin, i.e. half the time the coin will give heads (H) and half the time the coin will give tails (T).

Just as a coin, when tossed, can land heads or tails, so the photon when it hits a beam-splitter can with equal probability either go through or be reflected (i.e. reflection or transmission are equally likely). The experiment contains two sets of beam-splitters, one after the other, where the resultant beam from the first beam-splitter is recombined in the second beam-splitter. Imagine this experiment being analogous to tossing a coin twice. In such a case all four outcomes are equally likely, i.e. we have HH, HT, TH, TT. Interestingly, with the photon we only ever see one result, the other results never occur. So how does this trick work?

Imagine that the photon definitely passes through the first beam-splitter, it then enters the second beam-splitter via a specific path. If it does not pass through the beam-splitter, and is instead reflected, then it enters the second beam-splitter via a different path. On top of this, we don't know which of these scenarios happened, but we know one of them must be true. At the second beam-splitter, whichever path it enters from, the photon similarly passes through the beam-splitter or is reflected. Therefore after the two beam-splitters our photon has four possible scenarios: RR (reflected, reflected), PR (passes through, reflected), RP, or PP. As you can see, this is equivalent to the coin tossing scenarios where after two tosses we have similar results. You know the results of the coin toss are definitely one of the four options, but until you look you just don't know which one. Amazingly though, this is not what is observed when we use a real photon rather than a coin. The real photon always comes out the same way from the second beam-splitter. But surely this is impossible?

If the photon either definitely passed through or was reflected after the first beam-splitter, then we should expect the same after the second, i.e. 50/50 reflection or passing through. Given that this never happens means that we need to re-evaluate what happened at the first beam-splitter. Strangely the only conclusion that we can reach is that it must actually be reflected and pass through at the same time. So what I am really saying is that the photon can actually exist in two spatial locations at the same time. This is the only way two beam-splitters can be combined to give the same outcome, time and time again.

The theatre crew, like the founders of quantum mechanics, were completely baffled by this. But, the fact that a quantum object exists in many different locations at the same time is experimentally beyond any doubt. Since the birth of quantum mechanics there have been numerous different ways of confirming this. I went on to explain further evidence showing how a photon can exist in two different locations at the same time.

Suppose that after the first beam-splitter we slow down the photon if it is reflected, but don't do anything if it passes through. This slowing down can be done by inserting a standard piece of optical equipment called a half-wave plate (itself just a piece of crystal with certain desired properties) in between the first and second beam-splitters. Interestingly, just by slowing down the reflected component after the first beam-splitter, we can alter what happens at the second beam-splitter! If previously, without the half-wave plate, the photon regularly used to come out of the second beam-splitter always reflected, then now, with the half-wave plate inserted, we can make it always pass through. This is the crux of the difference between a quantum object, such as a photon, and something classical, such as a coin. If the photon was random in the same way as a coin, this would be impossible. There is nothing we could do after the first coin toss to make the second coin toss absolutely certain. Coin tosses are completely independent, they happen definitively and in order, and nothing that we do once the first coin toss has happened can possibly change its result. However, with the photon, the first action is indefinite and hence both outcomes are still possible and equally real.

Ultimately this experiment can only be explained if we consider that the photon does not fully reflect or pass through, but actually does both at the same time. This is simply the only explanation that allows for the result that we observe. But what exactly is the mechanism that exploits this indeterminism at the first beam-splitter to give a definitive outcome after the second beam-splitter? This mechanism is known as interference, where one outcome is amplified and the others are cancelled.

'But water waves can also interfere. Is this quantum interference too?' one of the theatre crew exclaimed. The answer is no. With classical waves, such as water waves, it is always one wave interfering with another, altogether different, wave; in the photon example, however, it is a single photon interfering with itself. This is an altogether different proposition and has no analogue in classical physics.

Note again that we are not describing just one freak experiment with photons: there have been thousands of other experiments confirming this and with different kinds of objects (e.g. electrons, nuclei, atoms, molecules) that confirm that any object can potentially exist in several different places at the same time. In fact these experiments are standard fare for undergraduate quantum courses all over the world. Further to this, we think that the evidence is conclusive enough to suggest that any object in this Universe can be in many different locations at the same time.

At this point, the theatre crew seemed to have had enough, 'Hold on, Vlatko, this is all well and good, but how come we never see the same person in two places at the same time?' Of course they were correct in asking this; everything is composed of atoms and molecules so surely we should see everything in many places at the same time?

The answer why we don't see this phenomenon lies in another aspect of quantum theory, namely that of measurement. Measurements affect and change the state of the system being measured and through measurements we force the system to adopt one of its many possible states that existed prior to measurement. For example, even if the hero of the play was able to exist in several countries at the same time, as soon as someone asks him 'which country are you in?', the hero would have to give a single definitive answer. Similarly, while I could exist in several places at the same time, as I am here talking to the actors, the fact that they were listening was a kind of measurement they were performing on me – an interaction that meant that I could not be in any other place at the time.

So our hero leading an instantaneous parallel life in a different country with a different family as a storyline is all well and good, but this is not what quantum theory tells us. Quantum theory tells us that as soon as our hero was to interact with anybody or anything, he would be forced to adopt a unique position. All objects in the Universe are capable of being in all possible states, unless they are forced by measurement to reduce to a subset of those states.

So how can it be that a photon interacting with a beam-splitter or the half-wave plate does not constitute a measurement? This brings us to the heart of the measurement problem in quantum mechanics. Some kinds of interaction seem to destroy the quantumness and give us a definitive answer, whilst others preserve the ability of physical objects to be in many different places at the same time. The rule is that if we need to know the exact value of some property of an object (e.g. spatial location, momentum, energy), then we have to destroy the quantumness to obtain it – otherwise we can leave the quantumness intact.

What is more common than us forcing the quantum object to lose its quantumness, is that it can also lose its quantumness naturally through its interaction with the environment. In fact every object is constantly fighting a battle with the environment. The environment always wants to know more about the quantum object and this is much like a measurement. This is one of the challenges physicists spend a lot of time thinking about and working on. Currently we cannot prevent an atom from strongly interacting with the environment for more than a few seconds (this is currently the most optimistic estimate and only applies to some properties of an atom). The point is that even within this time-frame, we are able to use the quantum indeterminism to do some pretty funky stuff, such as quantum computation and quantum cryptography.

This quantum indeterminism, being in many different states at the same time, is not just confined to the microscopic domain. It is also responsible for all kinds of amazing macroscopic effects that we see around us. It can be used to understand how electrical current can flow without any resistance whatsoever in a superconductor, how neutron stars manage to overcome gravity, how big spiders manage to climb vertical walls, and why you don't fall through your floor given that there is so much empty space between the atoms in your floor. The answers to all of these puzzling questions lie in an understanding of quantum theory.

As I continued my discussion with the theatre crew, I said: 'This indeterminism stuff is not so hard to grasp, yet it's so important. I don't understand why kids are not taught this in school'. In fact, here in the UK, even if you do physics at high school, quantum mechanics is not taught in any meaningful way. I personally think this is a shame, as the earlier you are exposed to this kind of thinking, the easier it becomes later on in life to accept that this is actually how reality is. Grown-ups are just too sceptical and inflexible and sometimes we find our limited imagination working against us. Since our discussions, I believe Jon is in agreement as he has taken the affirmative action to design another play to communicate quantum theory to junior high school kids. I look forward to these kids eventually turning up in research journals, telling us old timers how we have just missed the point, and that quantum theory is even far more powerful than we thought possible.

At this stage I felt that I was doing all the talking and that they were doing all the listening, so I asked them what their play was about. The beginning of the story introduces one of the main characters telling a story about a spy in the USA. Upon hearing the word 'spy', I interjected, and thought why not tell them about one of the most striking applications of quantum theory, that of cryptography. They might just be able to weave it into their play.

The science of cryptography, which studies how to make communications more secure and how to break communications, i.e. the art of code-making and code-breaking, has developed markedly with the use of quantum theoretic principles. It is one of the simplest and most effective applications of quantum theory to date.

The need to communicate secretly goes back to the dawn of civilization. The Spartans, who were regarded as the best Ancient Greek warriors, developed a very elaborate cryptographic method, using the scytale, some 400 years before the birth of Christ.

When the whole outcome of the battle could depend on some crucial information communicated between Spartan commanders it was important that the enemy could not get hold of it and use it to their advantage. The scytale was a long stick (the word itself means a baton) around which a strip of cloth was wrapped, following which a message was written on the cloth. The point was that when the cloth was unravelled, the letters of the message appeared to be completely scrambled. One needed the stick itself (or a stick of similar circumference), to decipher the original message; otherwise you could not be sure how the letters line up and what the original message was.

The Spartan method of encryption of messages is generally known as permutation. Permutation is a formal mathematical name for an orderly scrambling of the letters in a sentence to make it incomprehensible. Any type of scrambling will do, as long as it is structured and the receiving party knows how to unscramble it.

Another method of secret communication known to the Ancients is that of substitution. It was first used by Caesar, some 50 years before the birth of Christ. How did Caesar communicate secretly with his army leaders? The idea was very simple. He would write down the alphabet: ABCDEFGHIJK…, and then use the shifted sequence: DEFGHIJKLMN…instead. So whenever A appears in his sentence, he would replace it by D. Every B would be substituted by E, and so on. The last three letters of the alphabet would then be represented by ABC. The order 'ATTACK TOMORROW' would then be encoded as 'DWWDFN WRPRUURZ'. And it is clear that just looking at the raw encrypted message doesn't make a lot of sense to anyone.

So in each case unless we knew exactly how the alphabet was shifted or knew the circumference of the stick used to encode the message we wouldn't be able to decode the communication. Since ancient times substitution and permutation ciphers have evolved considerably from parchments and cloth and there are now some very complex computational encryption models based on these tricks which are practically unbreakable (that is, unless you have 1000-odd parallel computers or about 150 years handy). What is surprising is that though substitution and permutation are still two of the most popular tricks to encode secretly, they are inherently breakable. It is comforting to know that to break current models would take around 1000 parallel computers or 150 years of processing time. However, given that computational power is growing all the time, how long can these approaches really remain secure?

In fact, of the tens of thousands of cryptographic protocols known to us, the only unbreakable one, as proven by Shannon, is the 'one-time pad'. The one-time pad is a cryptographic algorithm that cannot be broken simply through extending our computational power ad infinitum. It has a much stronger basis and, if done correctly, no amount of computing power currently or in the future is going to help. In theory it is very simple and only requires four rules to be obeyed to obtain perfect security:

1 The two parties that wish to communicate securely must already share a secret key. This key is used to convert your message into an 'encoded' form and then de-convert it back to the original form. It can only be known to both parties in communication (but cannot be known to anyone else). Once the secret key is established then the two parties can choose to communicate securely using that key at any time in the future.

2 The key you use to encode your message must be completely random. If there is a way of predicting if the next digit is zero or one, the code can no longer be used as this leaves the protocol open to a structured attack.

3 The key must be as long as the message itself, i.e. the number of bits that make up the key is the same as the number of bits that make up the message. If it is shorter then the protocol will be left susceptible to attack.

4 The key must be used only once (again, repetitions leave the code open to attack).

This is all very well, and if we could do all of these we would have perfectly secure communication and many of the cryptographers around the world would be out of a job. But in reality the art and development of cryptography is alive and well because the one-time pad is so darned difficult to use. Running through the requirements, requirement 3 and 4 seem simple enough, requirement 2 raises some eyebrows – I mean what is a completely random key?, and require-ment 1 just seems plain ridiculous. What's the point of two parties already needing to share a secret key in order to communicate a secret message? Surely whatever method was used to securely share the secret key can also be used to share the secret message. This just turns out to be a bit of a 'Catch 22' – you need the secret key to communicate secretly, but you also need to communicate secretly to establish the secret key in the first place! The problem of perfectly secure communication (assuming we can satisfy requirement 2) then reduces to the problem of being able to share a secret key in the first place. This problem is known as the key distribution problem and is one of the main challenges in cryptography, to which standard computer science and engineering have no final answer.

This is where quantum theory comes in – not only does it help us with requirement 1, and hence the key distribution problem, it also helps us with requirement 2 – that the key should be completely randomly generated. The latter we will elaborate on in Chapter 10.

Quantum theory gives us a novel solution to the key distribution problem. By making use of the fact that any measurement to determine a state irrevocably changes the state we can tell when the state has been tampered with. This is what we discussed earlier, where each bit of a key could exist in several different states at any one time. Then, if an eavesdropper covertly tries to capture any information from the key, this interference acts like a measurement and hence forces the bit to adopt one state or the other. The sender and receiver can then analyse a subset of the key in order to determine whether there is any eavesdropping present. If not, then they use the key; otherwise they discard everything and start the process again. Of course, the eavesdropping detected may not have been due to the presence of a human agent, but may have been due to some kind of noise on the line. However, to be safe, we must assume the worst case.

This approach was discovered in the early 1980s by an American, Charles Bennett, and a Canadian, Giles Brassard, and has been implemented successfully on a wide variety of fronts. An alternative method proposed independently by Artur Ekert, an Oxford physicist, is also very popular. Currently the bit rate is not very high, but if you want to send a single message super-securely, then this could work for you. For example, in October 2007, a quantum cryptography system was tested by Nicolas Gisin during the Swiss National Elections in Geneva. The Geneva canton (jurisdiction) used a dedicated line for counting the ballots and communicating this information securely between the counting office and the central polling station. The message was short, so it required very few bits, but it was of the utmost importance, and the Swiss did not want to risk any tampering. This was a genuine application of quantum theory to facilitate a problem for which we have no solution classically.

This is all well and good, but now where does this leave us? The key question is whether and how, with the introduction of quantum theory, the concept of information is affected. Now, instead of having a definite outcome as we saw in the earlier chapters, quantum mechanics tells us that we can have several possible outcomes. In the previous chapter, when we talked about social ordering, information was seen as a binding force between different social units. Similarly, information here is also a binding force between different aspects of a quantum system.

One would think we were fairly safe in assuming that mutual information cannot ever exceed 100% (i.e. it cannot be better than perfect). If all children who go to good schools perform magnificently in life, we would say that there is 100% mutual information between good schools and successful life (whatever your definition of a successful life might be). You surely cannot share more than 100% of information. However, odd as it may sound, quantum systems can actually share more mutual information than 100%. Any theory of information needs to be able to handle this in order to fully describe reality; otherwise there will be parts of reality inaccessible to our understanding.

To explain how information can exceed 100%, imagine a simple two-state quantum system, such as the 'spin' of an electron. Using the metaphor of spinning tops, we can consider that electrons are like small spinning tops, each 'spinning' in its own way, depending on the external circumstances. An electron, just like the spinning top, can be considered to spin clockwise or counterclockwise in any given direction: horizontally, vertically, at 45 degrees, and so on. Astonishingly, if we measure the electron spin at two different times, then the mutual information between these measurements can actually exceed any value that we could had previously thought possible. In classical physics, spins at different times can be correlated in the horizontal or vertical direction, so that, if the first measurement of spin yields 'horizontal clockwise', so does the second measurement, in which case, the mutual information between them is 100%. Real electrons, on the other hand, behave quantum mechanically and their spin measurements could be correlated in the vertical direction at the same time as the horizontal direction (and all other directions!). This is because electrons can spin simultaneously in the clockwise and counterclockwise directions, something that no spinning top can do. In this case we can say that they may share up to 200% information. This super-correlation is known as quantum entanglement, or 'spooky action at a distance' as Einstein referred to it.

There is a particularly fruitful way of viewing mutual information between objects that is responsible for the common sense phrase 'The whole is at least as large as any of its parts'. Suppose you have two friends, Steve and Bryan, and each of them has a choice of what to do next. Steve could either continue with his current employment or quit and find another job. Bryan faces the same choices. If they are completely uncertain about their future, we would attribute one bit of information to each of them, since they each have a choice of two possibilities. Now it is clear that the information residing in the joint future of Steve and Bryan is at least as big as each of the individual pieces of information. This is easy to see by imagining that we found out that Steve left his job. So there is no more uncertainty about his future. However, we still don't know what Bryan will do. The uncertainty is now as large as one bit, and it all lies in Bryan's choice. This is precisely the same as the 'I'll have a drink, if you have a drink' example we gave earlier. It is interesting that though the causal link in the two scenarios is different, it still leads to the same mutual information. Moreover, mutual information does not depend on causality at all; it is quite reasonable that two parties that have no causal link whatsoever can still share 100% correlation.

We can now phrase this all in terms of entropy, the measure we saw earlier to reflect the degree of uncertainty in a system. The entropy of the whole system must (classically speaking) be at least as large as the entropy of any of its parts. In this sense it is useful to think of entropy in terms of an area. The area of the USA is at least as big as any of its states. Even if some states overlap and contain common regions this surely must always be the case. So if you are looking for someone in the USA, your uncertainty concerning their whereabouts must be at least as big as your uncertainty if you know they are somewhere in California. The uncertainty of their position in California cannot be larger than the uncertainty of their position in the whole of the USA, given that there are many additional places in the rest of the USA to hide.

Astonishingly, in quantum mechanics this sentence no longer holds. In the quantum version of this task it may be more difficult to find the person if you focus specifically on California than if you search the whole of the USA. Somehow, in the quantum version, the area of California turns out to be larger than the overall area of the USA!

So what is our evidence that quantum systems behave in such a manner? Well, now I guess you are asking if this is for real. It seems very counterintuitive. This is true and moreover we can do an experiment with, say, two spinning electrons, to test it. They are created in such a way that there is absolutely no uncertainty about their overall state, but, if we look at each of them individually, it looks like a complete mess. In other words, they need each other to fully describe their state. This does not happen classically; if one of the systems is in a mess then adding another system cannot decrease the amount of mess. Ultimately this goes back to the fact that the two electrons are really super-correlated in quantum physics, some might say like 'identical twins' – i.e. there is some additional mutual information 'beyond the norm'.

So if this is how reality really works, can we still describe it in terms of information? The answer is affirmative, but clearly the Shannon entropy is not enough. The problem with Shannon's information is that it always tells us that there is at least as much information in a whole as there is in any of its parts. But, as we discussed, this is not true for quantum systems. Information therefore requires a more generalized treatment than that proposed by Shannon.

The key change is to upgrade the (classical) notion of a bit to something called a quantum bit or 'qubit'. A qubit is a quantum system that, unlike a bit, can exist in any combination of the two states, zero and one. All other ingredients of Shannon's theory remain intact. To quantify information, instead of using the entropy of a bit, we take the entropy of a qubit. This was first done by another student of John Wheeler, Ben Schumacher, who also coined the term qubit. The change from bit to qubit, although it may appear trivial, actually has profound implications. Going back to the example of the whole being less than some of its parts, the quantum entropy of two correlated quantum systems can now indeed be smaller than the entropy of each of them individually.

The theory of quantum information is a superset of Shannon information in that it reduces to Shannon information in certain cases. The key thing that quantum information tells us about reality, beyond what we have already learnt through Shannon, is that there is a great deal of untapped potential in terms of what we can achieve in information processing. Quantum information theory is already being exploited to design a new order of super-fast computers, highly secure cryptographic systems and, believe it or not, to teleport objects across vast distances.

The reader may be worried that the upgrade to quantum information now suddenly invalidates all the conclusions from Part One of the book. Quite the opposite in fact is the case. We could have approached DNA from the perspective of quantum information but, as it is a macroscopic phenomenon, the results of using the classical theory of information appear to capture fully the most important function of DNA. The same can be said for any of the information processing systems in the first part of the book. All such systems could be re-derived from the perspective of quantum information theory, but as they are macroscopic systems, the approximation given by classical information theory is sufficient. This is not to say that quantum physics does not apply to macroscopic objects – on the contrary, it applies to all matter in the Universe. It's just that its predictions are much less distinct from conventional physics at this level – so naturally we tend not to bother with the additional complexity of using quantum physics when we feel we don't need to.

From the point of view of information we can summarize two of the most important features of quantum theory; first, that qubits can exist in a variety of different states at the same time; and, secondly, when we measure a qubit we reduce it to a classical result, i.e. we get a definitive outcome.

Both of the above features could be either positive or negative characteristics, depending on the application. Regarding the notion that qubits can exist in a variety of different states at the same time, a positive is that the qubit has a far more intricate structure than a classical bit. This gives more power and flexibility to quantum information processing than was possible with our classical theory of information. Conversely, in some cases, as in cryptography, if the receiver doesn't have precise details then he is unable to decipher the message without damaging the state. Likewise with the measurement. On the positive side, we have explained that in quantum cryptography, it allows us to detect eavesdroppers; however, on the negative side, it reduces a qubit to a bit, and therefore reduces our information processing ability.

The pros and cons of quantum information are best seen in the next chapter as we discuss quantum computing.

Key points

- Quantum physics started a hundred years ago, describing how small objects behave.
- Two mind-boggling features of quantum physics mark it apart from anything we have seen so far. One is the possibility of objects being in two or more different states at the same time. An atom can be here and there at the same time, for example. The other one is the intrinsic randomness in the behaviour of quantum systems. We simply cannot in most cases tell what a quantum system will do even when we know everything there is to know about it.
- Understanding that quantum physics is all about information helps us to develop real world applications to achieve a new order of communications.
- Quantum cryptography is one of the areas where quantum physics has demonstrated a new order of information processing, where we can now communicate more securely than we ever thought possible.
- Quantum cryptography is not just a theoretical construct; it has been successfully implemented over vast distances.

Surfing the Waves: Hyper-Fast Computers

Who hasn't heard of a computer? In a society entirely dominated by these transistor infested boxes there are probably only a few remaining isolated tribes in the Amazon or around the Kalahari that have not been affected. From organizing our finances, flying a plane, warming up food, controlling our heartbeat (for some), these devices are prevalent in each and every aspect of our society. Whether we are talking about personal computers, mainframe computers, or the embedded computers that we find in our mobile phones or microwave ovens, it is very hard to even imagine a world without them.

The term computer, however, means more than just your average Apple Mac or PC. A computer, at its most basic level, is any object that can take instructions, and perform computations based on those instructions. In this sense computation is not limited to a machine or mechanical apparatus; atomic physical phenomena or living organisms are also perfectly valid forms of computers (and in many cases far more powerful than what we can achieve through current models). We'll discuss alternative models of computation later in this chapter.

Computers come in a variety of shapes and sizes and some are not always identifiable as computers at all (would you consider your fridge a computer?). Some are capable of doing millions of calculations in a single second, while others may take long periods of time to do even the most simple calculations. But theoretically, anything one computer is capable of doing, another computer is also capable of doing. Given the right instructions, and sufficient memory, the computer found in your fridge could, for example, simulate Microsoft Windows. The fact that it might be ridiculous to waste time using the embedded computer in your fridge to do anything other than what it was designed for is irrelevant – the point is that it obeys the same model of computation as every other computer and can therefore – by hook or by crook – eventually achieve the same result.

This notion is based on what is now called the Church–Turing thesis (dating back to 1936), a hypothesis about the nature of mechanical calculation devices, such as electronic computers. Alan Turing and Alonzo Church introduced the formalization of an algorithm and a 'purely mechanical' model for computing, that all modern computers today obey. The thesis claims that any calculation that is at all possible to perform can in fact be performed by an algorithm running on their computational model (which assumes that sufficient time and storage space are available). This leads us to the notion of a universal computer, on which all modern computers are based.

The drive for miniaturization of computing technology cannot have gone unnoticed. In particular, computers have been getting smaller (and hence faster, as closer circuitry means less distance to travel) ever since the first computer was built by von Neumann in the 1940s. In the late 1950s the then chairman and co-founder of Intel, Gordon Moore, noticed a very interesting and remarkable trend: every two years or so computers tend to double in their speed and memory. Moore noted this trend in one of his reports and this has since become known as Moore's law.

But why has Moore's law been upheld in the last 50 years? It certainly is not a natural law as it depends on the existence of humans and is largely related to factors within our control. Perhaps instead it should therefore be known as 'Moore's trend' or 'Moore's observation'. Some people more cynical of the computing industry may even say that Moore's law has been correct because of Moore (CEO of Intel, remember). By noting the law, he effectively set the standard for other companies. Given that Intel have such a dominant position within the microchip industry, Moore's law is thus a bit of a self-fulfilling prophecy.

If technology continues to abide by Moore's law, then the continually shrinking size of circuitry packed onto silicon chips would eventually reach a point where individual elements would be no larger than a few atoms. Then what? Where do we go from there? How small and fast can computers be?

Surely, however, there is a natural limitation to this exponential growth. At the moment we are using about 100 electrons to encode one bit of information. But in about 10 years time, we may be using one electron to encode one bit of information. Can we go beyond this and what is the ultimate limit? If physics tells us anything, it's that we should never be too dogmatic about our conclusions. History is littered with 'no-go' statements which are later proven to 'go' (e.g. remember Lord Kelvin's statement that machines heavier than air cannot fly, said only 30 years before the Wright brothers proved him wrong). So even as we find the ultimate limit to computation, there is already inherent uncertainty as to how long this limit will hold.

To understand the ultimate limits we first need to understand what computers are all about. Well, it is simple: computers are all about processing bits. The computers we have at present use the laws of Boolean logic to shift, change, and reshuffle bits (i.e. zeros and ones). George Boole published his Boolean algebra in 1854 with a complete system that allowed computational processes to be mathematically modelled. Remember that Shannon used this in the construction of his communication process. The implementation of Boolean logic currently is in terms of transistors, however there are a variety of plausible alternatives. As we know, classical bits, by definition, exist in one of two different states at any given time – a zero or a one. With quantum mechanics, however, we are permitted to have a zero and a one at the same time present in one physical system. In fact, we are permitted to have an infinite range of states between zero and one – which we called a qubit. The number of states a qubit could occupy is infinite because in principle we can tweak the ratio of probabilities in which the states 0 and 1 occur to any desired accuracy. When with certainty we have either 0 or 1 then this reduces to the classical case.

Also, in line with Moore's trend, at the atomic scale the physical laws that govern the behaviour and properties of the transistor circuitry (which is based on semiconductor technology) are inherently quantum mechanical in nature, not classical. Thus the question of whether a new kind of computer could be devised based on the principles of quantum physics is therefore not a forced endeavour but a necessary and natural next step.

It is also interesting to note how transistors, which are the 'neurons' of all computers, work by exploiting properties of semiconductors. The explanation of how semiconductors function is entirely quantum mechanical in nature; it simply cannot be understood classically. Are we thus to conclude that classical physics cannot explain how classical computers work? Or are we to say that classical computers are, in fact, quantum computers?

The fact is that classical information processing can be a good approximation to reality at the macroscopic scale, and that sometimes the high level of detail that it offers is sufficient for everyday purposes. In fact it's not unlikely; had we not have been near the quantum limit now, where we are forced to start looking at computing with individual atoms, perhaps there would not have been even the same degree of motivation behind researching into quantum computers and the underlying quantum information theory.

Quantum computation, as it is, is an extremely exciting and rapidly growing field of investigation. An increasing number of researchers, with a whole spectrum of different backgrounds, ranging from physics, via computing sciences and information theory to mathematics and philosophy, are involved in researching properties of quantum-based computation.

That the laws of classical physics lead us to completely different constraints on information processing than computation based on quantum mechanics, was first realized by an American physicist, Richard Feynman. Later and somewhat independently, this idea was extended significantly by his British colleague, David Deutsch. Both being students of the remarkable John Wheeler, whom we met in Chapter 1, it is perhaps not surprising that they were driven by the same question of the fundamental link between physics and computation.

Two of the most successful applications of quantum computing are in the factorization of large numbers and searching a large database. The first problem is important, as much of modern day cryptography is based on the difficulty of factoring large prime numbers (we will discuss this later), and the second problem is important because *any problem* in Nature can be reduced to a search for the correct answer amongst several (or a few million) incorrect answers. Searches are so ubiquitous that they range from you searching for a file on your computer to a plant searching for a molecule in order to convert the Sun's energy to useful work (we'll discuss this later, too).

So why does quantum mechanics help here? Why can't we do these with our normal everyday computers? Well the point is that, yes we can, and we do use our computers for these purposes, but as the size of the prime factor or list to be searched grows, it takes longer and longer to get an answer. Quantum physics helps with these kinds of problems, because unlike a conventional computer which checks each possibility one at a time, quantum physics allows us to check multiple possibilities simultaneously.

Enter another couple of illustrious employees of Bell Laboratories, Peter Shor and Lov Grover (to add to Claude Shannon, whom we met earlier). Whilst Shannon was looking at the optimization of messages sent over a telephone wire, Shor, in 1992, was looking in more detail at the security of these messages.

Security is important in many aspects of life. Just as you want your credit card details to be secure when you are paying for something, governments and various companies want their documents to be securely stored and unavailable to the public or other governments or companies. As we discussed in Chapter 8, secrecy in the modern world is based on the notion that something is 'computationally secure', i.e. it is secure in the sense that to break the code would require an inordinate amount of computing time and power. For, example, it is very easy for computers to multiply two numbers. You can check it yourself. Take two one-hundred digit long numbers (they are huge, like for example the number 10000000000000 0000000000000000000000000000000000000000000000000000 00000000000000000000000000000000000000000) and ask a computer to multiply them together. This, the computer will be able to execute in a split second, and you'll hardly notice that it's taken any time at all.

On the other hand, finding factors of a large number is very difficult. This is because there are simply many possibilities to explore. Imagine the 100. What are its factors? Two times 50 is equal to 100. But so is 4 times 25. Or 5 times 20, or 10 times 10. The number of factors grows quickly and finding all of them presents a great difficulty for any current (classical) computer (it's exponentially slower than multiplying numbers in the first place).

How is it that a quantum computer can factorize efficiently? The explanation, first presented by Shor and now known as Shor's algorithm, is that a quantum computer, by exploiting the quantum principle of superpositions, can exist in many different states at the same time. Imagine a single computer in a superposition of being in many different spatial locations at the same time. In each of those locations you can configure the machine to divide your number by a different number to search for factors. And this is a massive, stupendous speed-up, since one quantum computer is now simultaneously performing all these divisions, one in each different spatial location. And, if one of them is successful – we have our factors!

Have you ever wondered why your PIN (personal identification number) is secure when you withdraw money from an ATM machine (automatic teller)? How come that neither the bank staff nor the bank managed know your PIN? Why do they not obtain it when you type it into the ATM and steal your money?

The reason is that the ATM machine performs the following operation. When you type in your PIN with the intention of withdrawing the money, this (usually a four to six digit number) gets multiplied by a huge (say a 500 digit) number. The resulting number (a number 504 digits long) is than checked by the bank. And if it is in the database, you will be allowed to proceed with your transaction.

But, and this is the crucial but, the bank cannot figure out your PIN from the 504-digit long number that they have in their database. It would simply take them a very long time – longer than the age of the Universe with current computers!

The punch-line of all this is that, using a quantum computer, we can factorize numbers very quickly. If we have a quantum computer with 10,000 quantum bits, we could factor a 500-digit number in a few seconds. And that would be the end of most current security!

Lov Grover, on the other hand, in 1996, was interested in an altogether different problem. Grover wanted to know how to design an efficient search algorithm using the mass parallelism offered by a quantum computer. His idea can be explained through the following example: suppose that someone gives you access to a library containing a lot of unsorted books. If you want to find a particular book, then you simply have to search through all the books until you find the one you are looking for. If there are a million books to go through and, if it takes a second to check each book, that could take a long time indeed (one million seconds is equal to about two weeks)! A quantum computer could speed things up greatly and would only take a thousand seconds (instead of a million) – which is about a couple of hours – and this is what Grover managed to prove.

In a list with four entries (two bits of information labelled 00, 01, 10, 11) we would normally require a maximum of three searches to find the right one. This is because you would have to look at each of the elements and, if you are unlucky, the first three elements will not be the ones you are looking for. Quantum search can, on the other hand, search a four-element quantum database in only one step. As the size of the database increases so does the quantum advantage.

Searching a database of four elements and finding the one that we want is similar to tossing two coins and observing the result of both. The first coin toss would correspond to asking which half of the database the desired element is in (i.e. is it in the top half or the bottom half), and the second coin toss would then give you the exact desired element (i.e. of the two elements left, is it one element or the other). It is an inescapable conclusion that in classical physics (hence classical computing) you would need a minimum of two coin tosses or two steps to uniquely distinguish between four outcomes.

Using quantum computers, however, we can do this with half the effort, in only one step. This quantum computation is analogous to the example of a single photon going through two beam-splitters. There we also had four different possibilities (RR, PR, RP, PP), but we only needed a single photon to generate a definite single outcome. The quantum property of superposition allows one photon to explore four different possibilities at the same time, and ultimately, through interference of the different paths, will compress them into a definite single outcome (i.e. the element we are searching for). This logic can be generalized so that a quantum computer could be designed to scan any number of database elements much faster than a classical computer.

The fact that the search problem is one of the problems in which quantum computers offer a speed-up is particularly interesting given that nearly any problem can be phrased in terms of a search algorithm. For example, even if we consider factorization, this can be rephrased in terms of a search algorithm in the sense that we are searching through all possible factors that would give us the desired answer. In this case there may be more than one answer, but conceptually the problem is the same. Also, the factorization example is a good one, as in this case the search algorithm is proven to be not as efficient as Shor's algorithm. This is as expected. The search algorithm is a general algorithm and is applicable for any search problem, whereas Shor's factorization algorithm is specifically geared for the problem of factorization and does so using hidden properties related to the structure of the problem. Therefore, to put things into context, whilst quantum computation does offer a significantly more powerful computational model than classical computation, it is not implausible that there are classical problems that can still be more efficiently solved with a classical algorithm than with the general quantum search algorithm.

The main limitation of quantum computation geared towards solving classical problems is that we ultimately have to make a measurement in order to extract the answer, given that the question we are asking requires a definite answer. Whilst this measurement is necessary, it is an intrinsically probabilistic process and there is always a finite probability that our answer may be wrong. In some cases, such as Shor's algorithm, it is straightforward to classically verify whether the answer is correct, and if it's not we run the quantum computer again until we have success. In others we need to find more creative solutions. The reader need not get too distracted with this limitation as it's not a major obstacle to quantum computing in practice. A far more serious inefficiency is the effect of environmental noise which is, in practice, very difficult to control. Any quantum computation that wants to be more efficient than its classical counterpart has to be able to deal with both of these issues.

In addition to the factorizing and searching quantum algorithms there are a host of other more specialized problems that quantum computers could do significantly better than their classical counterparts. However, probably the most fascinating application of a quantum computer lies in simulating complex physical systems. Some systems are so complex, our atmosphere for example, that they are extremely difficult to simulate using present computers and take an enormous amount of time. In fact, even far simpler systems, such as simulating a physical system with 20 atoms, are already extremely difficult for a normal computer. At the same time these problems are very important to us and we would like answers within our lifetime. Of course, we simulate these problems today on normal computers, but we are unable to simulate them in their entirety and compromise by simulating only a few key features in order to get an answer within a reasonable period of time. We'd all like better predictions of climate, not just day by day, but also in the long run. Being able to predict the climate more accurately is not only important to make our lives more pleasant; it could be crucial for our survival on Earth. And for this, we definitely need better understanding of the evolution of various weather patterns.

Simulating other systems with quantum computers is a subject in its infancy. It lies in such uncharted territory that we don't even know any of its borders or limitations. I personally think that this area will explode in the future, but its real power is difficult to fully grasp at this stage.

Interestingly for engineers and computer scientists, all this quantum weirdness, far from being a hindrance, can be utilized for technological purposes to construct quantum computers that fundamentally use this weirdness to run much faster than any current computer can. Thinking of computation as a process that maximizes mutual information between the output and the input – i.e. the question being asked, we can think of the speed of computation as the rate of establishing mutual information, i.e. the rate of build up of correlations between the output and the input. Furthermore the fact that qubits offer a higher degree of mutual information than is possible with bits, directly translates into the quantum speed-up that we see in Shor's and Grover's algorithms.

The good news is that small-scale quantum computers have even already been experimentally realized. These quantum computers, although operating on a very limited basis, have still been successful in demonstrating a degree of speed-up beyond anything possible with classical computers.

In terms of the make up of quantum computers, qubits could be encoded in atoms, subatomic particles, many-atom clusters, in light, or indeed in some combination of these. However, none of these at present offer a medium to store, say, 1000 qubits in a superposition state, for long enough to assist with more complex calculations.

How far away are we from fully fledged quantum computers? By this, I mean how far are we from an implementation that can hold thousands or millions of qubits in a superposition? In short, the answer is quite far; the current record (depending on how you view it) is somewhere between 10 and 15 qubits. However, there are many ways of encoding information quantum mechanically simply because there are many different systems that could be used to encode quantum information, e.g. ions in an ion trap, photons in a cavity, free photons, nuclear spins in nuclear magnetic resonance, electron spins in electroparamagnetic resonance, and a whole host of solid state devices (such as superconductors). We are not short of candidates for implementation!

Experimental physicists these days have an almost perfect control of atoms. They can engineer all sort of things, such as isolating a single atom inside a very small trap (on the order of a million times smaller than a metre). How do they know that a single atom is sitting there? They know, because they can shine some laser light into this small area and if the laser light is reflected inside, this then allows them to see this atom.

So what can we perform with 10–15 qubits?. Not much that our existing computers cannot do. We can, and we have, multiplied two numbers, 3 and 5 to obtain 15. So that's a start! Anyone can of course do that, but the significance of these kinds of experiments is to show that this can be done at the quantum level, which is a whole new game. Conventional computers currently need some 10,000 atoms to achieve multiplication of 3 and 5. So, quantum computers are in this sense more efficient. What is a shame is that no current quantum technology knows how to compute with anything more than 15 qubits.

The question is: why is it difficult to scale up quantum computers? The answer to this question is the same as to the question we asked in Chapter 8, i.e. why don't we see quantum effects at the macroscopic level? The difficulty, simply stated, is that we need large systems to be in many different states at the same time in order that they demonstrate quantum behaviour. But, the larger the system, the more ways there are for the particular information about the state to leak out into the environment. And as soon as any information leaks out, superpositions, and with them quantum computers, are destroyed. The process of information leaking out can also be considered as the environment (i.e. what is outside of the system) performing a measurement to gather information about the state of the system. As we know from our discussion on cryptography this inevitably damages the quantum state in question. This process is known in technical parlance as *decoherence*. So basically, it is difficult to make large-scale quantum computers for the same reason that we don't see people in multiple places at the same time (unless after a few drinks).

An intriguing analogy here is with crime. Statistics indicate that most successful robbers are those who operate by themselves (women apparently do this more frequently than men and are therefore much more successful robbers). The point to this is that the more people you have in your gang, the bigger the chances that one of them will betray you for whatever reason. It makes intuitive sense somehow that the more people that possess the same information, the higher the probability that this information will get out. Well, you can think that it is exactly the same principle as with atoms – the more atoms there are in a superposition, the harder it is to stop one of them decohering to the environment.

This all sounds like bad news for quantum computers. Is it going to be impossible to make reliable large-scale quantum computers? The answer fortunately is 'no'. Unlike the perfect crime, we can in fact have perfect quantum computers. How is this? The trick to achieve this is the same that makes DNA replication faithful, or at least faithful enough for life to propagate. The solution is redundancy!

If you want to have 100 useful quantum bits in your computer, then you have to build a 1000-qubit large quantum computer. This means that for every one qubit, you have an additional nine copies of that qubit, so, if the occasional one decoheres, you can still make a majority decision over the remaining nine. Quantum superpositions can thus be protected, and whilst there are more optimal ways of adding redundancy, this in essence is the key concept.

As I write these lines (summer 2008) we are already moving towards producing quantum computational microchips. Furthermore, this is happening right now and right here in Singapore (I say this as I am sitting down and sipping a double espresso in Spinelli's on the National University campus). On these microchips, experimental physicists are trying to integrate components involving super-cold atoms and coherent photons in small numbers. Micro-engineering has progressed so much in the last 20 years that it is amazing to see how much stuff can be packed onto a very small area of a two-dimensional chip. Making a fully quantum computational unit can easily take some 5–10 years to complete, but so much ingenuity has gone into the research here that the possibility of large quantum computers is a very real prospect!

As with living systems, the battle to build a quantum computer is ultimately a battle against entropy. The lower the overall entropy of an arbitrary physical system the higher the chances that its constituent atoms may be entangled. Typical atoms useful for quantum computation usually need to be at a temperature close to absolute zero (around 1 billionth of a kelvin). Furthermore, given that the temperature in the rest of the laboratory is 300 billion times higher than that, it's a constant battle. This is a kind of Maxwell's demon scenario, where a process needs to reduce the entropy of the system in order to get some useful information processing done.

A striking piece of evidence to show how quantum effects can be seen in some macroscopic objects was demonstrated by Syantani Ghosh and her colleagues in 2003. Ghosh showed that quantum superpositions between many atoms exist in a piece of salt involving billions of atoms and at temperatures of a few millikelvin. This was a huge shock because it showed that quantum phenomena, whose power was thought to be confined to the infinitesimal world of subatomic particles, can produce effects that remain measurable on macroscopic scales.

This discovery was so momentous, it was difficult for anyone to believe. In a paper I submitted in 2000 to *Nature*, the premier scientific journal, I made similar predictions and was duly laughed off the court. The fact that the prediction has now been borne out by Ghosh's experiment catapults the mystery of quantum information into the wider arena. *Nature* kindly sent me Ghosh's paper as they thought I would be pleased to know that one of my predictions turned out to be correct (it doesn't happen too often). Even better still, since Ghosh's results, a number of other results have demonstrated similar effects in other materials, some of them being at much higher temperatures than this (even, startlingly, at room temperature!).

We are now realizing that advanced quantum effects are much more ubiquitous in macroscopic systems and this gives us hope that one day we may find that Nature has already provided us with a quantum computer and all that is left for us to do is to program it. After all, Nature has already invented many tricks before us humans. Radar and error correction by redundancy are just two out of many such tricks used by living systems. Might it therefore be that somewhere there is a living system which harnesses the speed-up advantages of quantum computation? Even better, perhaps quantum computation is so ubiquitous that it takes place in every living cell.

There is continuing evidence that more and more natural processes must be based on quantum principles in order to function as they do. For example, we consider the process of photosynthesis, which is one of the key natural processes for maintaining life on Earth.

The reader will recall from Chapter 5 that all living beings are like thermodynamical engines similar to Maxwell's demons, whose crucial task is to battle the natural tendency to increase disorder. Life does that by absorbing highly disordered energy coming from the Sun and converting it to a more ordered and useful form. Photosynthesis is the name of the mechanism by which plants absorb, store, and use this light energy from the Sun. This energy is converted to an ordered form and used for the functioning of the cells. Recent fascinating experiments by Graham Fleming and his group at the University of Berkeley, California, suggest that quantum effects do matter in photosynthesis. Furthermore, they point out a close connection between the photosynthetic energy transfer and Grover's optimal quantum search algorithm. In other words, plants are so much more efficient than what we expected that maybe there is some underlying quantum information processing contributing. Biological plant efficiency is super-high, around 98% of radiation that hits the leaf does get stored efficiently. The best man-made photocells, on the other hand, can barely achieve a 20% efficiency. So there is an enormous difference; how do plants do it?

The complete answer is not entirely clear but the overall picture is this. When sunlight hits a surface that is not designed to carefully absorb and store it, the energy is usually dissipated to heat within the surface, or reflected. Either way, it is lost as far as any subsequent useful work is concerned. The dissipation within the surface happens because each atom in the surface acts independently of other atoms. When radiation is absorbed in this incoherent way, then all its useful properties vanish. What is needed is for the atoms and molecules in the surface to act in unison. And this is a feat that all green plants manage to achieve.

In order to understand how this happens, it is helpful to think of each molecule in a plant as a database element in Grover's search algorithm. All molecules vibrate as they interact with one another. When they are hit by light, they change their vibration and dynamics and what ensues implements Grover's dynamics with the energy ending up in the most stable configuration at the end (and this configuration is the database element that Grover's algorithm is meant to identify).

Admittedly, Fleming's experiments have been performed at low temperature (77 kelvin, while plants normally operate at 300 kelvin). Therefore, it is not entirely clear if any quantum effects can survive at the higher temperature. However, even the fact that there exists a real possibility that quantum computation has been implemented by living systems is a very exciting and growing area of research.

But things could be even grander. The most fascinating sugges-tion of all is that quantum theory may be necessary in the very basis of life itself. You will remember that in Chapter 4 I said that the infor-mation to replicate life is encoded into four different elements, which are just molecules labelled by A, C, T, and G. If we think of this as a very simple database, with four records, the task of a DNA replicator computer is to scan an arbitrary DNA strand and match the correct record from the database to each of the molecules. The database logic is that A is always paired with T, and C is always paired with G. This simulates the replication process in our cells, as each molecule on one DNA strand finds its companion on the new strand. In this way we can consider DNA replication analogous to a four-element database search problem.

An Indian physicist called Arvind asked the question in the late 1990s as to why Nature uses two bits (four molecules) to encode life. Wouldn't it be simpler just to use one bit (i.e. just two molecules)? This, as you may recall, was one of our big questions in Chapter 4. It seems odd to use two bits given that we know that everything can be done in terms of single bits and it seems intuitively simpler to use single bits, so why has Nature made the extra effort? Shouldn't Nature, starting from nothing, find it easier to stumble upon a single-bit encoding, than a two-bit encoding? Arvind's answer is simple: It may be easier to come up with single bits, but there is a bigger quantum advantage in search when we have two quantum bits. With two quantum bits Grover's quantum search algorithm finds the solution in just one step. Therefore, if we optimize for the speed of replication, then two quantum bits allow for more efficient information processing than one classical bit in the same number of steps. So maybe this is what Nature was thinking.

Arvind's suggestion is very smart, but there is a crucial issue here: can DNA actually be a quantum computer? DNA is a macromolecule so it's not clear how this could be the case and how molecules could exist in many different states at the same time. Whether DNA is based on classical or quantum computing is, remarkably, not actually known. I say 'remarkably' because DNA has been studied extensively for the last 60 years. We therefore have to be patient and leave this question open for the time being.

The picture that seems to be emerging is that larger and larger systems may be capable of exhibiting quantum effects under certain conditions. We are not sure, for example, if and how we can generalize this. Perhaps everything would exhibit quantum effects if we know how to look at them in the right way. So is it the case that any complex piece of matter or energy can potentially, under the right external conditions, be used as a quantum computer? If we take a step further and couple this to the Church–Turing thesis of the universal computing machine, what this could imply is that any piece of the Universe can simulate, in a more or less efficient manner, any other piece of the Universe. Perhaps even reality itself can be seen as the product of a complex multilayered quantum computation. This is, of course, a huge leap of faith but it will be the essence of our discussion in Chapter 12.

Key points

- Quantum computers force a higher order of information processing than we can currently achieve. They are the smallest and fastest gadgets that the laws of physics currently allow us to construct.
- They can solve some important problems for us that conventional computers cannot. Two instances are the factorization problem and the search problem. The former is used in various security protocols, the latter in many optimization techniques.
- Quantum computers are not a distant dream. They are being built at present in many laboratories in the world.
- Quantum effects are experimentally being seen in macroscopic objects such as pieces of solids and organic molecules in living systems.

Children of the Aimless Chance: Randomness versus Determinism

In our search for the ultimate law, P, that allows us to encode the whole of reality we have come across a very fundamental obstacle. As Deutsch argued, P cannot be all-encompassing, simply because it cannot explain its own origins. We need a law more fundamental than P, from which P can be derived. But then this more fundamental law also needs to come from somewhere. This is like the metaphor of the painter in the lunatic asylum, who is trying to paint a picture of the garden he is sitting in. He can never find a way to completely include himself in the picture and gets caught in an infinite regression.

Does this mean we can never understand the whole of reality? Maybe so, given that any postulate that we start from needs its own explanation. Any law that underlies reality ultimately needs an a priori law. This puts us in a bit of a 'Catch 22' situation. So, are we resigned to failure or is there a way out? Is there some fundamental level at which events have no a priori causes and we can break the infinite regression?

What does it mean for an event to have no a priori cause? This means that, even with all prior knowledge, we cannot infer that this event will take place. Furthermore, if there were genuinely acausal events in this Universe, this would imply a fundamentally random element of reality that cannot be reduced to anything deterministic.

This is a hugely controversial area, with various proponents of religion, science, and philosophy all having a quite contrasting set of views on this. Often people get very emotional over this question, as it has profound implications for us as human beings. Could it be that some events just don't have first causes? The British philosopher Bertrand Russell thought so. In Russell's famous debate with Reverend Copleston on the origin of the world, Copleston thought everything must have a cause, and therefore the world has a cause – and this cause is ultimately God himself. Copleston asks: 'But your general point then, Lord Russell, is that it's illegitimate even to ask the question of the cause of the world?', to which Russell replies: 'Yes, that's my position', and the whole debate reaches a dead-end as the two diametrically opposing views refused to compromise.

Interestingly, the quantum theory of information adds a novel twist to this millennia-old question of determinism versus randomness. This question is not just important for our understanding of reality, it is also important to us on a very personal level. The answer to this impacts on whether there is any room for genuinely free action in an ordered and structured Universe like ours. If the laws of reality govern everything, then even our actions would be subjugated and determined by them. This of course leaves us no room for the human element we call 'free will' – a property that we strongly feel distinguishes us from non-living matter (and other animals). It is also seen as the basis of our consciousness.

Most of us in the West feel that determinism cannot completely govern reality because we are certain that we have free will, though what exactly this amounts to is far from uncontroversial. For the sake of discussion, let us define free will as the capacity for persons to control their actions in a manner not imposed by previous events, i.e. as containing some element of randomness as well as some element of determinism. So, if we accept the notion that we do indeed have free will, then we are already, in some sense, entertaining the idea that there may be a random element to reality (obviously not all elements of reality can be random, because this too would exclude any role for free will).

This still raises an intriguing question, simply because either of the two possible answers – 'yes we do have free will', or 'no we don't'– seem to lead to a contradiction. For example, suppose you answer with 'yes, we have free will'. How would you demonstrate the validity of this statement? You would need to act in a way that would not be predetermined by anything. But how can this ever be, when whatever you do, can, in fact, be predetermined by something? To further qualify this argument, say you decide to act out of character, e.g. having an introverted personality you decide to start a conversation with a complete stranger on the street. But, the very fact that you decided to act contrary to your usual predisposition, seems itself to be fully predetermined. It is simply so by the fact that you determined that you would act out of character to prove free will. In this case perhaps in trying to prove free will, you are more likely to demonstrate that actually you have none. Your emotions could have been controlled by some outside factors which lead you to the conclusion that you must act out of character. If they were, all you are really trying to do is deterministically fight determinism, which is by definition a deterministic process!

The considerable difficulty in demonstrating free will conclusively lead us to postulate that perhaps we cannot have it. But this answer feels completely contrary to the whole of human psychology. Can nothing good that I do be attributed to me? Is it all predetermined by my genes or my history or my parents or social order or the rest of the Universe? Even worse, we tend to reward people for doing good deeds and punish them for bad ones. This would seem to be completely misconceived if humans did indeed not have any free will. How can you punish someone for doing something when they are not free to do otherwise? Is our whole moral and judicial system based on the illusion of free will? This just feels wrong, although there is no logical reason why free will is necessary.

That we seem to have no way of proving that there is free will was poetically stated by the famous biologist Thomas Henry Huxley: 'What proof is there that brutes are other than a superior race of marionettes, which eat without pleasure, cry without pain, desire nothing, know nothing, and only simulate intelligence?'

Free will lies somewhere between randomness and determinism which seem to be at the opposite extremes in reality. It's clear that neither pure randomness or pure determinism would leave any room for free will. If the world is completely random, then by definition we have no control over what will happen, and if the world is completely deterministic we also similarly have no control over what will happen as it would all be pre-scripted. So you are stuck between a rock and a hard place.

But are randomness and determinism actually opposite extremes when it comes to defining reality? Are they mutually exclusive, meaning that they cannot both exist within the same framework? Our latest model of physics, quantum theory, suggests that there is a way of combining the two. Every quantum event is fundamentally random, yet we find that large objects behave deterministically. How can this be?

The answer is that sometimes when we combine many random things, a more predictable outcome can emerge. This may seem paradoxical at first (shouldn't lots of random things give you something even more random?) but this is not necessarily the case.

Imagine you toss a coin 100 times. Each of the outcomes is very uncertain and you'd have difficulty predicting it. Your success would not be better than 50%, which amounts to just random guessing. But imagine that instead of betting on individual outcomes, you in fact bet on the number of heads and tails when 100 tosses have finished. This is much easier, since 50 heads and 50 tails is what you would expect if the coin was fair. So, each toss is random, but overall a predictable pattern emerges.

In physics we encounter this all the time. Every atom in a magnet can be thought of as a mini-magnet but its behaviour is very erratic. A single atom's magnetic axis is impossible to predict, unless we apply a strong external influence to align it. However, even without this external influence, all these randomly aligned atoms can together still produce a magnet with a clearly defined north and south. Therefore, a deterministic event can emerge out of a random one.

Randomness at the microscopic level therefore does not always propagate to the macroscopic level. It is perfectly plausible that in spite of quantum mechanics being our most accurate description of Nature, the world of large objects – the one that matters to us humans most in our everyday lives – is fully deterministic. This would imply that even though the world is random at the microscopic level, there is still no free will at the macroscopic level.

But what exactly does it mean to be random? We think of tossing a coin and observing the outcome (heads or tails) as a random process. It is random because before we observe the outcome it has an equal chance of being heads or tails. And it is very difficult to predict the outcome. But what if we knew all about the coin? These would include its weight, exactly how it was tossed, and any relevant properties of the air around it. Then Newton's laws tell us that we should be able to predict the outcome of each toss. Therefore, based on classical physics, randomness is superficial. There is no fundamental randomness once we have all the information.

Quantum physics, with its peculiar blend of randomness and determinism, obviously makes our debate about these topics far richer. Uncertainty in classical physics, unlike in quantum physics, is not fundamental but simply arises out of our ignorance of certain facts. George Boole expresses this idea clearly: 'Probability is expectation founded upon partial knowledge. A perfect acquaintance with all the circumstances affecting the occurrence of an event would change expectation into certainty, and leave neither room nor demand for a theory of probabilities.' In quantum physics however this statement does not hold. One of the most fundamental and defining features of quantum theory is that even when we have all information about a system, the outcome is still probabilistic. According to quantum theory, reality can be fundamentally random and not just apparently random (i.e. where we may be missing some information).

Here is a very simple experiment that is a quantum mechanical equivalent of tossing a coin. Imagine a photon, which as we said is a particle of light, encountering a beam-splitter. Recall that a beam-splitter is just a mirror that has a certain silver coating on it, so that we can tune the probability with which the photon is reflected or transmitted. Let us say that these probabilities are made equal, so that we have the exact equivalent of a fair coin.

Just as a coin, when tossed, can land heads or tails, so the photon impinging on a beam-splitter can go through, or be reflected. And, as far as all experiments indicate, the beam-splitter experiment is completely random. Each time a photon is sent to a beam-splitter, we can in no way predict its path subsequently. Reflection or transmission are equally likely and occur randomly.

But now I would like to make a distinction between a coin and a photon. The behaviour of a coin is random, not fundamentally, but because it is unpredictable. The behaviour of a photon is not just unpredictable, it is genuinely random. What can this possibly mean?

The coin is governed by the laws of classical physics. If we knew the exact initial conditions of the coin toss, namely the speed and angle, then we would in principle be able to fully predict the outcome. But it may take us a very long time to be able to compute the outcome. Perhaps it takes years or longer to do this. Nevertheless, the equations that govern the coin dynamics are fully deterministic and in principle can be solved to tell us the outcome. Therefore, the coin toss only appears random. It is really deterministic, but difficult to predict.

Now let's consider the situation with the photon and the beam-splitter. Here the equations are quantum, and they are also deterministic. The indeterminism of quantum mechanics manifests itself only through deliberate or non-deliberate (environmental) measurements. When the system is undisturbed by either, we have a clear and well-defined deterministic view of it as described by the Schrödinger equation. And what this equation tells us is that after the beam-splitter, the photon has both gone through and has been reflected. Yes. Both possibilities occur in reality. The photon is now in two places at once. It is behind the beam-splitter, having gone through, and in front of it, having been reflected. Unlike the coin toss, it is very simple to solve quantum equations to reach this conclusion. But the conclusion seems paradoxical!

And what happened to randomness? It comes in through the backdoor. Imagine that we now want to check where the photon is, i.e. we want to make a deliberate measurement. If we put detectors behind and in front of the beam-splitter they will then record the presence or absence of the photon. Let us say that the detection, when it happens, is amplified to a loud click. What happens in the actual experiments? The detectors click randomly.

There is absolutely no way to tell which detector will click in each particular run. This randomness is genuine; it is not just unpredictable as a coin toss is. Why? How do we know that the equation telling us that the photon is on both sides before it is measured cannot be supplemented by another deterministic law to tell us what happens when the photon is detected?

The reason why we know that the 'coin toss type' randomness is not involved is as follows. Suppose that instead of measuring the photon after the first beam-splitter we put another beam-splitter after the first one instead. What happens then? If the behaviour of the photon was classical, then it would be deterministic at both beam-splitters, but the outcome would be even more difficult to predict. We would therefore expect the photon to have random outcomes at the second beam-splitter too. Half of the time it should be detected in front of it and half of the time behind it.

But this is not what happens in the lab. In an actual experiment, the photon now deterministically always ends up behind the second beam-splitter. So, two quantum random processes have, in a way, cancelled out to give something deterministic. And this is something that cannot happen classically. Imagine that you toss a fair coin twice and you find out that it always comes out heads both times. But when you toss it once sometimes it is heads and sometimes it is tails. This could, of course, never happen. But in the case of a photon it does. This is why we believe that there is something very different in the behaviour of the photon to what any classical system would do.

So what have we learnt? We've learnt that all elementary quantum events are in fact fundamentally random. But this of course doesn't mean that a reality built on these random events must necessarily be random. We have seen how randomness and determinism can co-exist and moreover determinism can emerge from random origins. This gives us some insight into the origins and nature of reality.

There is a beautiful quantum protocol that illustrates how randomness and determinism can work hand in hand to produce a stunning outcome. Think *Star Trek* and the teleportation chamber. Does anyone actually believe that teleportation is possible? No? Well you'd better believe it! Teleportation 'dematerializes' an object positioned at a location A only to make it reappear at a distant location B at some later time. OK, so quantum teleportation differs slightly, because we are not teleporting the whole object but just its quantum information from particle A to particle B, but the principle is the same (after all, the whole thesis of this book is that we are all just bits of information).

As all quantum particles are indistinguishable anyway, this amounts to 'real' teleportation. What I mean by this is that all electrons, for example, have identical properties (mass, charge) and the only distinguishing feature is how they spin. Given that all other properties of two electrons are the same, if we also manage to encode one electron spin into another, then we can consider this a successful transfer of quantum information between the two. In other words, the second electron becomes an identical copy of the first, and the two are now indistinguishable. This is true for all particles such as protons, atoms, and so on.

One way of performing teleportation (and certainly the way portrayed in various science fiction movies, e.g. *The Fly*) is first to learn all the properties of that object (i.e. get all the information about what makes that object) and then send this information as a classical string of data to location B where the object is then re-created.

One problem with this proposal is that, if we have a single electron and we don't know its spin, we cannot determine it because this would require us to make a measurement thereby invariably damaging the original quantum information in the spin. So it would seem that the laws of quantum mechanics prohibit teleportation of a single quantum system (unless we know its state in advance).

However, it will turn out that there is no need to learn the state of the system in order to teleport it. All we need to do is use mutual quantum information, of the same sort that exists in a quantum computer. This provides super-correlation between locations A and B (technically known as quantum entanglement), so that quantum information can be transferred. Even though the quantum measurement driving the teleportation is still random, this can be overcome by sending some auxiliary classical information from A to B. The actual measurement outcome at A would be communicated to B and this could even be done over a standard telephone line. After the teleportation is completed, the original state of the particle at A is destroyed. This then illustrates how teleportation at the quantum level can actually be performed – and this in spite of the intrinsic randomness.

Currently we can teleport only individual atoms and photons over only a couple of metres. The basic principle has therefore been experimentally verified, first by Anton Zeilinger's research group at the University of Vienna, and independently by Francesco de Martini's group at the University of Rome, but the question still remains if we can ever use this to teleport larger objects and ultimately humans. With humans things could get even more complicated. If we faithfully teleported every atom in your body, would that necessarily be the same as teleporting you? In other words, is your body all there is to you? The answer to this is that we really have no idea!

Now, in a more general sense, in order to distinguish superficial (classical) from real (quantum) randomness we really need a clearer measure in order to quantify the difference. Interestingly enough, one such measure does exist, and although it is closely related to the Shannon entropy, it has quite a different motivation. To understand why the Shannon entropy may itself not be so useful as a measure of randomness, consider the following example.

A typical random process as indicated is a coin toss. It generates a sequence of heads (H) and tails (T). Typically, after 10 tosses, say, we expect to have something looking like HHTHTTTHTH. We much less expect to see something like HHHHHHHHHH. The first sequence just looks like one that should come out of a random source, like a coin: The second looks too ordered to be genuinely random. We might also think of the second one as being less likely. But this is a mistake.

Here we therefore have a problem. Since heads and tails have an equal probability of one-half, any sequence of heads and tails is as likely to come up as any other sequence (people playing the lottery rarely appreciate this fact: 1, 2, 3, 4, 5, 6 is as likely as 2, 3, 17, 30, 41, 45). It is true that a sequence with half heads and half tails is more likely than the one with all heads. But this is because there are more such sequences: there is only one sequence involving all heads, while there are about a thousand sequences with 50–50 heads/tails (HHHHHTTTTT, or HTHTHTHTHT, or TTTHHTTHHH, and so on).

Any particular random sequence is as likely as any other. But, still, we have a strong feeling that while the sequence HHTHTT-THTH looks random, the sequence HHHHHHHHHH, appears to be very orderly. Their probability seems not to be able to capture this basic difference since both are equally likely (or unlikely; the probability is about one in a thousand for either). And this is why Shannon's entropy fails to quantify randomness, as it is based only on probabilities.

The measure used to solve the problem of quantifying randomness was ultimately introduced by a Russian mathematician, Andrey Kolmogorov, in the late 1950s. His solution is in some sense very operational – or, better still, physical. His solution to randomness was as follows. How random a sequence is depends on how difficult it is to produce this sequence. Produce it with what? Answer: a computer.

Imagine writing a program for your computer to generate a sequence of heads and tails. The sequence of all heads (HHHHHH-HHHH) only needs one instruction: 'print 10 heads'. But a sequence HTHHHTTHTH cannot be generated as easily. In fact the program probably just needs to say 'print the sequence HTHHHTTHTH'. Therefore, when we have a random looking sequence, the program to generate it is at least as long as the sequence itself (in our case it is longer since it requires additional instructions, e.g. 'initialize computer, print the sequence, end program', etc. but this will be a marginal difference when the string is very long anyway). However, when something is orderly, the program can be much shorter. The quantity that tells us by how much orderly things can be compressed to shorter programs is known as Kolmogorov's complexity.

Are there any problems with this definition? If we are not careful it seems that different computers could give us different estimates of how random something is. Fortunately we remind ourselves of Turing's notion of the universal computer, a computer that can simulate any other. So, to avoid such issues, Kolmogorov suggested calculating the Kolmogorov complexity using this universal computer.

Using Kolmogorov's ideas, we can re-examine the difference between classical superficial randomness and quantum fundamental randomness. We said that the sequence of tosses of a coin of the type HHHTTTHTTHH is random because it cannot be given a shorter program to generate it. Imagine having all the information that fully captures this coin toss (e.g. the relative weight difference between the head and tail face of the coin, speed of rotation, height thrown to, etc.). With all this information fixed, to the correct level of precision, we could write a program that could generate the sequence for any number of coin tosses. This is because classical physics is fully deterministic. The length of this program in classical physics would necessarily be shorter than doing the coin tosses themselves (for a large number of coin tosses). In the quantum case, this would not be so. Given the problem of predicting the sequence of photon clicks in a detector (our quantum coin toss) the program used to describe the sequence could only describe it by actually running each experiment individually. Therefore, the program would have to be at least the same length as the sequence of detector clicks it is trying to describe. There is no short-cut!

An intriguing possibility now emerges. Can we apply Kolmogorov's logic to understand the origins of reality? Can we say that our laws that describe reality are exactly this – an attempt to understand the dynamics of the Universe and reduce its complexity? In this way, the laws themselves (in physics, biology, economics, sociology) could all be seen as short programs describing the makeup of reality. For example, in physics, rather than running each and every experiment (i.e. writing the whole program) we can write a shortened program which only uses the current laws of physics to predict the outcome of each experiment. This latter program would hence be significantly more efficient than the program that actually runs all the experiments individually. In this way we can consider our duty as scientists to find the shortest program that represents reality. So let us look into how this reduction in complexity is achieved in science.

To understand what exactly is at stake, we first need to understand the logic of science, physics in particular. This was something that the philosopher Karl Popper devoted his life to. His way of understanding science will be the key to understanding reality in my final chapter. So let us summarize it here.

When Popper was growing up in the 1920s, physics was reaching its peak, but some other disciplines started to emerge, which are now known as social sciences. Sigmund Freud was pioneering psychology (through psychoanalysis) and sociology and political sciences were also emerging. While some people would be happy to call social sciences by the name of 'soft' sciences (as opposed to hard, fact-based science, such as physics), Popper was altogether very concerned with attaching the name science to anything like psychoanalysis.

His principal aim was to devise a criterion for calling something a science in the first place. The crucial idea that occurred to him is the following. While it is easy to falsify a physical theory (just do an experiment whose results clearly contradict the theory, and quantum experiments were a shining example of falsifying classical physics in Popper's time), it is not so easy to falsify a psychological theory.

How many times have you heard the remark that someone is not confident because their mother did not love them? Then again, you also hear that others are confident precisely because their mothers did not love them and they had to rely on themselves more. And so here is the problem. The theory 'his mother did not love him' seems to be able to explain too much. So much so that it can be used to justify two diametrically opposing facts, someone being confident and another person not being confident. This means that such a theory can never be falsified, or shown to be wrong, in practice.

The famous eighteenth-century Scottish philosopher David Hume was particularly bothered by non-falsifiability of some claims (psychoanalysis didn't exist in his time, he was much more bothered by philosophy and religion). He phrased it as follows. Any number of white swans we see in the world cannot prove to us the conjecture that 'all swans are white'. However, a sight of one single black swan is enough to destroy it. And so it is with science. Newton's physics had been tested for 200 years and always found to be correct. But one test in the late nineteenth century – in the shape of black-body radiation – was enough to destroy it. The black body was the black swan of physics, destroying the hypothesis that all physics obeys classical physics. Of course this does not mean that classical physics is of no use to us whatsoever, it just means that there must now be a new theory (i.e. quantum mechanics) that takes into account classical physics plus its black swan. In contrast to a concrete statement that can be falsified, e.g. 'all swans in this river are white', stands a statement like 'God works in mysterious ways' – I mean, how the bleep are you going to disprove that?

How can we therefore be sure about anything that comes from science? We cannot. But, rather than this being a problem, Popper thought that this was the whole point of science! Namely, a theory is only genuine if there is a way of falsifying it! If under no circumstances can you disprove your theory (i.e. create an experiment to rule it out), then this theory is worthless as far as knowledge is concerned because you can never test it. Popper therefore turned a seemingly negative feature of science (the fact that any theory could be proven wrong) into its most fundamental and necessary feature. It is through centuries of refutations, falsifying (i.e. refuting), and improving on theories (i.e. conjecturing), that science has progressed to where it is today.

Let us see if we can interpret Popper's logic within the context of information theory. Once a theory is established by a few experiments, we then start to gain more confidence in its validity (though we may be, and usually are, ultimately proven wrong). As a result we attribute a higher chance to the theory passing the next test. If the theory passes it, the principles of information would say that this is a low-information event. The reason is that the higher the probability of an event, the smaller the surprise when the event happens.

As our confidence grows, the probability we attribute to the theory being falsified becomes smaller and smaller. In the current experiments testing quantum mechanics, not many in the physics community expect quantum mechanics to fail. But this is precisely why it would be a great shock if it did. So, falsifying something usually carries far more information – both emotional and physical – than confirming it.

Any information in physics comes from the scientific method Popper called 'conjectures and refutations'. But even this method can be seen as a form of information processing. This will result in a rigorous and well-defined statement of Occam's razor. We can think of scientific theories as programs run on a universal computer, with the output being the result of whatever experiment we are trying to model. We say that our theory is powerful, if we can compress all sorts of observations into very few equations. The more we are able to compress, the better we believe we can understand something. Because then from very few laws we can generate the whole of reality.

Occam's razor just says that if there are many theories that explain something, then we should choose the shortest one as the correct one. The shortest description of Nature that generates all possible observations is to be preferred over a very long description. To quote Leibniz, whom we met in relation to one of the proofs of the existence of God, 'God has chosen that which is the most simple in hypotheses and the most rich in phenomena'. This statement would imply that the information in the Universe is highly compressible into a few simple laws.

But now we face an interesting question. Any theory we come up with will be finite, namely it will contain a fixed (hopefully small) set of rules. And this means – as was first fully realized within information theory by Gregory Chaitin, an American mathematician – that it can only produce a finite set of results. In other words, there will be many experimental outcomes that could not be compressed within the theory. And this effectively implies that they are random. This was also realized by Leibniz who stated: 'But when a rule is extremely complex, that which conforms to it passes as random'. This perfectly encapsulates the Kolmogorov view of randomness: when the rule is as complicated as the outcome it needs to produce, then the outcome must be seen as complex or, in other words, random.

Following this logic quantum randomness can be encapsulated in two principles, as first argued by the Italian physicist Carlo Rovelli. One of them is borrowed from classical information and simply states that the most elementary quantum system cannot hold more than one bit of information. This is almost self-evident as by definition a bit is the smallest unit of information. The second principle is that we can always obtain new information. This principle, when combined with the first, captures the fundamental randomness we see in quantum events. The only way we can obtain new information when we seemingly have all information, is if this new information is random. Can it be that this is just a restatement of the fact that a finite number of axioms can only lead to a finite number of outcomes? If this was so, the implications would be amazing.

There is still a school of thought that views randomness in quantum theory as due to its incompleteness, i.e. due to our lack of knowledge of a more detailed deterministic underlying theory. However, if we view the growth of our knowledge in physics through compression, it may suggest to us that randomness is inherent in the

Universe, and therefore it must be part of any physical description of reality. Randomness could simply be there because our description of reality is always (by construction) finite and anything requiring more information than that would appear to be random (since our description could not predict it).

This would mean that randomness in quantum physics is far from unexpected – in fact according to this logic it is actually essential. Furthermore, it would mean that whatever theory – if any – superseded quantum physics, it would still have to contain some random features. This is a very profound conclusion. Given that physics is always evolving, having fundamental randomness sets a serious constraint on any new theory.

Having the potential to be wrong is what Popper identified as the key aspect of scientific knowledge. However, this should also be true of any other form of knowledge (philosophical, psychological, religious, historical, artistic, you name it). In this respect useful scientific knowledge is the same as the financial gain in betting or stock markets. If there is no risk then there is nothing to gain – i.e. no free lunch! It's not just scientific knowledge or economic profit that grows in this way, any useful information within whatever context you can describe, always grows in this way.

Science is therefore just a form of betting on future outcomes. The idea of representing our uncertainty about the Universe through gambling was, in fact, already suggested by the famous German philosopher Immanuel Kant in his *Critique of Pure Reason*, as long ago as 1781. Kant equated betting with the pragmatic belief in the validity of our theories. The logarithmic scoring rule (so central to Shannon) is, in a sense, a practical implementation of this philosophical suggestion.

The interplay between randomness and determinism in fact propagates throughout the book. For Popper, randomness lies in making conjectures in science, and determinism in refuting these conjectures by deliberate experimentation. This was, according to him, the only way to secure information about the world, i.e. knowledge. But in other elements of reality, useful information also emerges in exactly the same way.

For example, consider the process of evolution of biological information. Biologists mainly think of information processing in living systems as created through evolution. Evolution has two components, one is the random mutation in the genetic code and the second is deterministic natural selection by the environment of this new feature. In this way, evolution of biological information is seen to emerge analogously to creating any useful scientific knowledge.

The same is true in economics, where the central goal is to understand and predict market behaviour. Whether it is economic policy, or a simple financial investment decision, any conjectured strategy will be played out in the market, and therefore falsified or confirmed.

We have also argued that social dynamics is another form of information processing. And furthermore the level of development of a society may be viewed as synonymous with its capacity to process information. Randomness appears between elements of a society, whereas through interaction with one another, at the collective level we see deterministic characteristics through all sorts of phase transitions that societies undergo. It is probably not an accident that societies that have applied the method of conjectures and refutations more vigorously have had the opportunity to develop faster.

In physics, randomness was seen to be crucial in the concept of heat, and in fact the whole Universe was seen as evolving towards the state of maximal entropy (maximal randomness or disorder). The deterministic part of this process was to use available information to design schemes to extract useful work efficiently. The whole of thermodynamics can be seen as a battle between deliberate Maxwell's demons, trying to extract order from disorder, and natural randomizing processes.

Everywhere we look we see underlying bits of information. Furthermore this information always obeys the same evolution through randomness and determinism, independently of context. So can a combination of randomness and determinism produce all information and everything else we see around us?

Key points

- Randomness and determinism together can be seen to underlie every aspect of reality.
- This is linked to the age-old question of free will; a question that has entertained us since the days of Ancient Greece.
- Through a series of conjectures and refutations we can now see how knowledge evolves.
- Quantum mechanics opens the door to genuine randomness (i.e. events which, at their most fundamental level, have no underlying cause).
- Kolmogorov captured the essence of randomness in that a collection of outcomes from a random process cannot be generated in any simpler way than by actually running that process (i.e. you just have to suck it and see).

PART THREE

The first part of the book was all about how information under-pins a number of everyday processes. Information is as critical to the biological propagation of life as it is to structuring a stock portfolio or getting useful work out of random energy.

The second part of the book showed us that there is more to infor-mation than meets the eye. When our underlying model of physics is adjusted from a classical view to a quantum one, we find that our theory of information is also adjusted and becomes even more insightful than before. The key advantage of quantum information is that a quantum bit can exist in several states at the same time, the flip-side of this being its inherent degree of randomness. This can sometimes deter us when we want to use quantum information to perform useful work, but it also presents some opportunities that we can exploit to our advantage (recall that in quantum cryptog-raphy, randomness was used to eliminate eavesdropping).

More importantly, for any information to become more complex or higher quality (e.g. when we gain more knowledge about the Universe) the genuinely random element is key. We need to keep coming up with new information through this random element and then have a process that deterministically eliminates that which is incorrect or not required. It is this deterministic element that allows us to form useful knowledge.

The first two parts of the book unify a number of seemingly unrelated ideas through the common language of information processing, but the question still remains: where does even this information come from? This takes us right back to what we asked right at the beginning of this book, namely the question of *creation ex nihilo*.

All the theories and experiments presented in the first two parts are beyond reasonable doubt; these are all peer reviewed and well accepted facts. In Part Three of the book we move into more speculative and uncharted territory.

Sand Reckoning: Whose Information is It, Anyway?

In Chapter 9 we discussed the idea of a universal Turing machine. This machine is capable of simulating any other machine given sufficient time and energy. For example, we discussed how your fridge microprocessor could be programmed to run Microsoft Windows, then we described Moore's logic, that computers are becoming faster and smaller. Therefore, one day, a single atom may be able to simulate fully what a present day PC can do.

This leads us to the fascinating possibility that every little constituent of our Universe may be able to simulate any other, given enough time and energy. The Universe therefore consists of a great number of little universal quantum computers. But this surely makes the Universe itself the largest quantum computer. So how powerful is our largest quantum computer? How many bits, how many computational steps? What is the total amount of information that the computer can hold?

Since our view is that everything in reality is composed of information, it would be useful to know how much information there is in total and whether this total amount is growing or shrinking. The Second Law already tells us that the physical entropy in the Universe is always increasing. Since physical entropy has the same form as

Shannon's information, the Second Law also tells us that the information content of the Universe can only ever increase too. But what does this mean for us? If we consider our objective to be a full understanding of the Universe then we have to accept that the finish line is always moving further and further away from us.

We define our reality through the laws and principles that we establish from the information that we gather. Quantum mechanics, for example, gives us a very different reality to what classical mechanics told us. In the Stone Age, the caveman's perception of reality and what was possible was also markedly different from what Newton would have understood. In this way we process information from the Universe to create our reality. We can think of the Universe as a large balloon, within which there is a smaller balloon, our reality. Our reality is based on our knowledge of the Universe (via the laws through which we define it) and as we improve our understanding of the Universe, through conjectures and refutations and evolutions of our laws and principles, the smaller balloon expands to fill the larger balloon. So is the rate at which the Universe keeps surprising us greater than the rate at which we can evolve our reality? In other words, will we ever understand our whole Universe?

Here Popperian logic comes to our aid. The very logic of conjectures and refutations, which is at the root of how we construct our understanding of reality, tells us that we cannot answer this question. Only if we know that the Universe can no longer surprise us, that there is no new physical theory superseding that which we currently have, can we be sure that one day we might be able to understand our whole Universe. But how do we know that no new event in the Universe will ever surprise us and cause us to change our view of reality? The answer is that we don't. We just don't know whether one day we will be surprised. This is the ultimate unknown unknown. Even though we can never know whether we will know everything, this does not prevent us from knowing how much there is to know with the current understanding. So how much information is there in the Universe? Where do we even start with such an absurd calculation? Interestingly we are not the first ones to tackle this issue and far greater minds before mine have grappled with this question.

One of the greatest human minds in history is Archimedes of Syracuse (circa. 287–212 BC). Archimedes made enormous contributions to astronomy, mathematics, engineering, and philosophy, and had a reputation at the time as being not only theoretically very astute but also very practical. In fact, he was regularly employed by the state for his ingenious ideas especially when they were in a pickle and didn't know who else to turn to. In one story, which saw Syracuse under attack from warring ships, his idea was to focus the Sun's energy using large curved mirrors, to burn enemy sails before the ships docked. This according to folklore was one of several times he had saved the city.

Uncompromising in his pursuit of science, Archimedes died as he lived. The last words attributed to Archimedes are 'do not disturb my circles' and it is said he made this remark to a soldier just before he was slain. Whilst the soldier originally came to summon him, Archimedes' complete apathy to the soldier's concerns, and total focus on his work, resulted in a tragic end to an otherwise spectacular life.

One piece of his research was commissioned by the Syracusian monarch, King Gelos II, and resulted in what is considered the first research paper ever. The task was to calculate the number of grains of sand that could fill the Universe. Not the kind of task you would set to any ordinary man. It is certainly not clear what use King Gelos II ever made of this, or whether he ever used it for anything other than fanciful conversation.

Sand was the smallest thing known at that time so it was natural to phrase questions in terms of volumes of sand. With reference to the heliocentric model of Aristarchus of Samos (circa. 310–230 BC),

Archimedes reasoned that the Universe was spherical and that the ratio of the diameter of the Universe to the diameter of the orbit of the Earth around the Sun equalled the ratio of the diameter of the orbit of the Earth around the Sun to the diameter of the Earth. You could see how this calculation was not everybody's cup of tea. In order to obtain an upper bound, Archimedes used overestimates of his data.

Firstly, to even begin to describe numbers of such epic proportions he needed to extend the current Greek numbering system. In practice, up until then they never had a language for such large numbers, but he needed it now. For example, Archimedes introduced the word 'myriad' to represent the number we would now call 10,000. Following from that a myriad myriads would be 100 million (i.e. 10,000 times 10,000), and so on. However, the real difficulty came as he had to make assumptions about the size of the Universe, using a very limited knowledge of astronomy (by our standards). Here is what he reasoned:

1 That the perimeter of the Earth was no bigger than 300 myriad stadia (roughly 50,000 kilometres – this is very close to the actual perimeter).

2 That the Moon was no larger than the Earth (it's actually much smaller) and that the Sun was no more than about 30 times larger than the Moon (this is a huge underestimate).

3 That the angular diameter of the Sun, as seen from the Earth, was greater than approximately half a degree (this is correct and not a bad estimate either).

With these assumptions, Archimedes then calculated that the diameter of the Universe was no more than 10 to the power of 14 stadia (i.e. in modern terms two light-years), and that it would require no more than 10^{63} (one followed by 63 zeros) grains of sand to fill it. In Archimedes' estimate, if you think of every point in space as a bit, in the sense that it either contains a grain of sand or not, then the number of bits according to him is 2 to the power of 10 to the power of 63, i.e. 2 to the power of 10^{63}. This is a phenomenally large number, and later we'll compare it with the equivalent number generated more recently, some two millennia after Archimedes took the first stab.

Archimedes, of course, never had the level of understanding of the world that we do now. So the question is how much more accurate can we be with a two thousand year head-start.

Over the last two thousand years we have seen Popper's method working tirelessly to produce ever better approximations of reality. Scientists would come up with conjectures on how to describe elements of reality in the shortest and simplest way and then they would perform observations and experiments to test their models (conjectures and refutations). As the models are refuted, as new information or understanding comes to light, they are discarded and new models arise from their ashes to take their place. So far we have been looking at biological, computational, social, and economic aspects of reality. What we have seen is that information is a natural way in which to unify these seemingly different disciplines. As discussed throughout this book, at its heart, the nature of information processing depends entirely on the laws of physics. So to calculate the amount of information in the Universe it is natural that we resort to our best understanding of reality to date.

The two theories which encapsulate our current best understanding of reality are quantum physics and gravity. There are, of course, other theories (biological, social, economic, and so on) that have a legitimate claim to this and we have given them an equal seat at Calvino's table. However, it is generally regarded that quantum theory and gravity are the most fundamental descriptions available to us. In the same way that Kolmogorov viewed the information content or complexity of a message in terms of the shortest program used to describe it, we currently view the information content of reality in terms of quantum physics and gravity – which are our shortest programs used to describe reality.

We have already seen in Chapter 10 how quantum theory can be understood in terms of information, so now can we also do the same for gravity?

Gravity is quite distinct from quantum theory. Whilst the effects of quantum theory can be felt at the microscopic and macroscopic level, with effects becoming less influential for large bodies, gravity works the other way around: gravity dominates large bodies (e.g. planets) and becomes less influential for microscopic bodies. No current experiment can detect gravity between two atoms, no matter how close they are.

These two theories may seem to have their place at opposite ends of the spectrum, but they are, in fact, intricately related. Finding a single unified theory connecting quantum physics and gravity has been seen as the holy grail of physics for some time now and is one of the 'bleeding edge' areas of physics. With tongue in cheek I will argue how gravity can be described as a consequence of quantum information (expounding on a view that is extremely controversial). This will, I believe, be the strongest indictment to date that quantum information does indeed provide the underlying description of reality.

The modern view of gravity, through Einstein's general relativity, is to see it as a curvature of space and time. In everyday language you may be more aware of it as a universal force of attraction, like when you throw a ball into the air and it comes back down. Einstein's view is, however, the most general and accurate description of gravity. In this view, time and space are inseparable and both curved interdependently. To visualize this, imagine a simple example where your bed represents space-time. As things are placed on the bed they create impressions by indenting the surface. If you put a football onto your bed it might make a slight impression; if you put yourself on to the bed this would make a much bigger impression. If you are sitting on the bed and you put the ball sufficiently near to you, it will be pulled in by your impression. In exactly the same way, all bodies attract one another in space-time because they create indentations in the space-time fabric which then propagate and affect each other. Understanding the curvature of space-time (or your bed as you sit on it) is therefore the key to describing the effects of gravity.

Any curvature in space-time necessarily means that distances and time intervals both become dependent on the mass of the object curving the space-time. For example, the time for a person closer to Earth runs faster than for someone further from Earth (assuming this person is not close to some other massive object) simply because the person closer to the Earth is more affected by the Earth's gravitational pull; in other words, the curvature of space-time is greater closer to the Earth. Given that we want to explain gravity in terms of quantum information, the question then is what is the connection between the geometry of the curvature and the concept of information (or entropy)? Amazingly enough, the answer lies in that quirky property we met in the last chapter, quantum mutual information.

We have met the concept of entropy in several chapters already. This concept is synonymous with the information content of a message or system. The higher the entropy of a system the more information it carries. In the first part of the book we used entropy for many different purposes; it quantified the capacity of a channel, the disorder in any physical system, the profit from a bet in terms of the risk taken, and social interconnectedness. For the purposes of the following discussion we should think of entropy as physical entropy, quantifying disorder in physical systems.

There is a very interesting relationship between the uncertainty within a certain system – as measured by its entropy – and its size. Suppose that we look at an object enclosed within a certain boundary. How complex would we expect it to be? A natural answer would be that the entropy of the object depends on its size, in particular its volume. Say that a molecule contains one million atoms arranged in a ball. If each atom has a certain entropy associated with it then it seems natural to expect that the total entropy is just the number of atoms, times the entropy of each atom. Therefore, the total entropy would scale as molecular volume.

Interestingly, it turns out that at low temperatures (such as the current state of the Universe), the entropy usually scales with the area and not the volume of an object (and as we know from elementary maths, the area of any object is necessarily always smaller than its volume). If you think of a ball-shaped molecule, the volume is made up of all the atoms on its surface plus the atoms inside. Therefore, if we are now asking what the maximum entropy of the ball-shaped molecule is, we might say it is proportional to the total number atoms in the ball (i.e. the volume), given that each atom should independently be able to contribute to the overall uncertainty. However, and rather surprisingly, what quantum theory is telling us is that, no – entropy is actually proportional to the total number of atoms on the surface (i.e. a significantly smaller ratio).

So why is there this difference between what seems logical and what quantum theory tells us? To understand this we have to look into quantum theory again, and specifically towards the nature of quantum mutual information. Recall that quantum mutual information is a form of super-correlation between different objects and that this super-correlation is fundamental to the difference between quantum and classical information processing (e.g. as we see in quantum computation).

Suppose that we divide the total Universe into two, the system, such as the molecule above, and the rest – which is everything outside of the molecule. Now, the quantum mutual information between the molecule and the rest is simply equal to the entropy of the molecule. But, quantum mutual information is not at all a property of the molecule, it can only be referenced as a joint property, i.e. a quantum correlation between objects. In this case it is a joint property between the molecule and the rest of the Universe. Therefore, it logically follows that the degree of quantum mutual information between these two must be proportional to something that is common to both, in this case the boundary – i.e. the surface area of the molecule!

This is a very profound conclusion. We think of entropy as the information content of an object. The fact that this information content is not within the object, but lies on its surface area, seems surprising to say the least! What this means is that the information content of anything does not reside in the object itself, but is a relational property of the object in connection with the rest of the Universe.

This also implies another important result. It is in fact a possibility which will be explored later. Within this formalism it is entirely possible for the Universe to have zero information content, whereas subsets of the Universe may have some information. The Universe is not correlated to anything outside of the Universe (by definition). However there are parts of the Universe that are correlated to each other. As soon as we partition the Universe into two or more distinct regions we begin to generate information, and this information is equal to the area of the partition, not to the size of the regions. The very act of partitioning, dividing, and pigeonholing necessarily increases information, as you cut through any parts that may be correlated.

We can present this conclusion pictorially as follows. Imagine that atoms inside the molecule are connected to atoms in the Universe via a series of ribbons. A limit on the number of ribbons that we can connect to the Universe is constrained by the surface area of the molecule (i.e. how many ribbons can we get through the surface of the molecule – as it is only of finite size). The information shared between the molecule and the Universe can be seen as proportional to the number of ribbons connecting the two. And this is logically why information scales as the area of the surface of the molecule.

The physicist Leonard Susskind proposed to call the relationship between entropy and area, the holographic principle. Holography has traditionally been part of optics and is the study of how to faithfully encode three-dimensional images onto photographic two-dimensional films. This is typically done by illuminating the object with laser light, which gets reflected off the object and the reflection is then recorded onto photographic film. When the plate is subsequently illuminated, a three-dimensional image of the object appears where once the real object stood. This phenomenon is probably familiar to the reader from its countless usage in magazines, stickers, toys, and science fiction films.

Optical holography was invented by Denis Gabor in the 1960s working in the labs at Imperial College in London (one of which became my office some 40 years later). He received the Nobel Prize for his ideas in 1971 (interestingly, the year I was born). He used holography to construct an optical version of Maxwell's demon (an idea which I also researched during my PhD). The whole surprise about his discovery is that he showed that two dimensions were sufficient to store all information about three dimensions, for which he duly received the Nobel Prize (sorry – though I thought hard but there's unfortunately no connection between us on this one!).

It's easy to see how two dimensions are recorded but where does the third dimension come from? It is this third dimension that allows us to see a hologram in three dimensions. The answer to this lies in the relational properties of light, known as interference. Going back to the experimental setup, light carries an internal clock, and when the light reflected from the object interferes with the light directly hitting the two-dimensional photographic film then an interference pattern is produced where the timing from the clock acts as the third dimension. This means that when you look at a hologram, you see the standard two-dimensional image, but you are also seeing light reflected back to you at slightly different times and this is what gives you the perception of a three-dimensional image.

Susskind suggested that we should not be surprised that information (entropy) scales with surface area, but rather we should elevate this to a point of principle. By this he means that this principle should be correct for anything in the Universe (anything that carries energy – e.g. matter, light). Furthermore, the key property behind this was quantum mutual information which we now see as being between anything on one side of the object and whatever is on the other side. Now we have all the pieces to derive gravity from this logic.

Einstein's equation in general relativity describes the effect of energy-mass on the geometrical structure of four-dimensional space-time. His equation says, to paraphrase John Wheeler, that matter tells space-time how to curve, while space-time (when curved) instructs matter how to move (e.g. the Earth moves around the Sun because the Sun curves space-time significantly). Can the energy-curvature relationship that encapsulates gravity be derived from quantum information theory?

An ingenious argument was given in the mid-1990s by Ted Jacobson to support the answer 'yes' and we now have all the ingredients to recreate it. So far we have already discussed how the thermodynamical entropy is proportional to the geometry of the system. It is well known in thermodynamics that the entropy of a system multiplied by its temperature is the same as the energy of that system. Therefore a larger mass, which represents larger energy (based on the mass–energy equivalence), will imply a larger curvature in space-time.

A simple energy conservation statement between entropy and energy becomes Einstein's gravitational equation, relating mass to curvature. In this case entropy encapsulates geometry. A more massive object therefore, according to thermodynamics, produces larger entropy. However, we saw that the entropy is also related to the surface area surrounding the mass, according to the holographic principle that we have just discussed. Therefore, the more massive the object, the larger the indentation of the surrounding area.

It is very helpful to illustrate this with an example. Take space-time without any mass or energy (i.e. an empty Universe). Now divide it into two by taking a sheet of light, shining it directly through the middle (whatever that means in an empty Universe). This light is unaffected by anything else since the Universe is empty. Now imagine introducing a massive object on one side (massive here simply meaning an object with a large mass). From thermodynamics, this changes the entropy, which from holographic principles affects the area that the light travels, which will have to be bent now to take into account the change in geometry. This, in fact, was how general relativity was first tested by Arthur Eddington in 1919. He confirmed that the apparent change in the position of a star followed exactly the prediction by Einstein in his work on general relativity. Here the light from the star was being curved *en route* to us via the gravitational pull of the Sun – hence the star appears as if it had changed position.

Interestingly the same idea can be applied to the detection of massive dark objects such as black holes. How can you see a black hole when by definition it is black and it does not emit any light? What the black hole does have, however, is a huge gravitational force, and we can use this fact to 'see' it. Whilst we cannot observe the black hole directly we can observe the impact of its gravitational force on matter and especially light around it. In particular the light of any star directly behind a black hole, instead of being dispersed in the usual manner, is instead highly focused as it gets caught up in the gravitational pull of the black hole. From Earth we observe the light from a star becoming significantly more intense than normal, before it settles back down to its normal level of light intensity. This change of intensity can be explained by a black hole passing in front of the star. In technical terms the effect is known as gravitational lensing.

Information, as measured by entropy, is now seen to underpin both quantum mechanics and gravity. In quantum mechanics, the entropy of a system is finite but we can always generate more entropy (which implies randomness). The fact that this quantum entropy is proportional to the area can then be coupled to the First Law of thermodynamics, which states that energy is conserved, to infer the equations of gravity. It is very interesting to note that quantum physics and gravity are frequently viewed as incompatible. However, this argument would suggest that, far from it, they are actually intimately related (which is why Jacobson's paper caused a lot of excitement).

We have already said that some aspects of this argument are speculative. However, what we can conclude from the whole discussion is that gravity does not add anything novel to the nature of information processing. All the theory required already exists through application of quantum principles. Even if the details of the argument are not correct, still the properties of quantum information are the same with or without gravity.

So with this in mind let's return to the question of how much information can maximally be squeezed into the total Universe as we know it. We have already said that information is proportional to area, and how exactly it is proportional has been estimated by the Israeli physicist Jacob Bekenstein. His relationship, known as the Bekenstein bound, is simply stated as follows: the number of bits that can be packed into any system is at most 10^{44} bits of information times the system's mass in kilograms and its maximum length in metres (the square of this length is the system's area). As an aside we note that Bekenstein's work on black hole entropy prompted the British physicist Stephen Hawking to conclude that (after all) black hole are not as black as they seem. They emit the so-called Hawking radiation, whose ultimate origin is quantum.

It is amazing that to calculate something as profound as the information carrying capacity of any object, out of its infinitely many possible properties, we only need two: area and mass. As a practical application, this easily allows us to calculate the information carrying capacity of our heads. Say that a typical head is 20 centimetres in diameter and weighs 5 kilograms. That means that a typical human head can store 10 to the power of 44 bits of information. Compare this to the best current computers which are still only of the order of 10 to the power of 14 bits of information. We therefore need 10^{30} to get the equivalent information carrying ability of a human head!

For us, the task now is to apply the Bekenstein bound to calculate the total number of bits in the Universe, which, if you recall, was our original motivation. Astronomers have already given us a rough estimate of the Universe's size and weight, say 15 billion light-years in diameter and a mass of about 10 to the power of 42 kilograms (ironically, this coincides with the 'forty two' from *Hitchhiker's Guide to the Galaxy*). When you plug this information into the Bekenstein formula, the capacity of the Universe ends up being on the order of 10^{100} bits of information. This is a stupendously large number, but ultimately it is not infinite. (In fact mathematicians will argue that it is still closer to 0 than it is to infinity!) Also it's worth noting that Archimedes estimated 10 to the power 63 grains of sand in the Universe. If, as before, we take a grain of sand as analogous to a bit of information, then this is a pretty good guess of the Universe's information carrying capacity from someone who lived more than two millennia ago.

Since we have been equating the Universe to a quantum computer, it would also be applicable to talk about the processing speed of our Universe. This can be estimated immediately from Bekenstein's bound. If you take the age of the Universe as 10 to the power of 17 seconds and the fact that the Universe has generated 10^{100} bits (these are our current estimates), then we can say that the total capacity for information processing is about 10^{90} per second. Comparing this to a modern computer (your everyday Pentium 4 – whose processing capacity is not more than 10^{10} bits per second) we can see that we would need 10^{80} such computers to simulate the Universe. This is 10 followed by 80 zeros. Therefore, if we had to rely only on our computers to understand the Universe we would not get very far! This is an amazing indication of the power of the human mind!

For comparison, at the other end of the spectrum, lie small objects such as atoms and atomic nuclei. A hydrogen atom, according to

Bekenstein, can encode about four million bits, while, a proton can encode only about 40 bits (simply because it is so much smaller than the atom itself). If we were skilful enough (and we are currently far away from this) we could run a quantum computation with just one hydrogen atom that would factorize a 1000-digit number, something we said was extremely difficult for any current computer.

How confident are we that this number of bits that we have calculated is the total number of bits of information in the Universe? Popper has already told us that you can view science as a machine that compresses bits in the Universe into laws, and these laws in turn are then used to generate reality. So how do we know whether tomorrow we will observe an experiment that will change the compression and give us a new law? The answer is that we don't! For a theory to have stood for 200 years and to then be refuted by a single experiment is standard fare for scientific progress. In the same way that quantum information has superseded classical information, it is likely that in the future we may progress to a new order of information processing, based on additional elements of reality unknown or not fully understood by us at this time.

So can we push this method of conjectures and refutations to the extreme? Can we present a consistent view of reality without even worrying about what the final theory would be?

Key points

- The entropy of any system is proportional to the surface area of that system. This is known as the holographic principle and it is a consequence of quantum mutual information.
- Using the holographic principle, we can estimate the number of bits in the Universe as well as the number of elementary units of information processing that it can hold.

- Interestingly a similar calculation was performed by Archimedes some 2500 years ago, when he tried to estimate the number of grains of sand in the Universe (and his answer was not so far off).
- The power of the Universe as a quantum computer is finite but way beyond anything we can currently imagine or have any idea how to use.

Destruction ab Toto:
Nothing from Something

The main view promoted by this book is that underlying many different aspects of reality is some form of information processing. The theory of information started rather innocently, as the result of a very specific question that Shannon considered, which was how to maximize the capacity of communication between two users. Shannon showed that all we need is to associate a probability to an event, and defined a metric that allowed you to quantify the information content of that event. Interestingly, because of its simplicity and intuitiveness, Shannon's views have been successfully applied to many other problems. We can view biological information through Shannon's theory as a communication in time (where the objective of natural selection is to propagate the gene pool into the future). But it is not only that communications and biology are trying to optimize information. In physics, systems arrange themselves so that entropy is maximized, and this entropy is quantified in the same way as Shannon's information. We encounter the same form of information in other phenomena. Financial speculation is also governed by the same concept of entropy, and optimizing your profit is the same problem as optimizing your channel capacity. In social theory, society is governed by its interconnectedness or correlation and this correlation itself is quantified by Shannon's entropy.

Underlying all these phenomena was the classical Boolean logic where events had clear outcomes, either yes or no, on or off, and so on. In our most accurate description of reality, given by quantum theory, we know that bits of information are an approximation to a much more precise concept of qubits. Qubits, unlike bits, can exist in a multitude of states, any combination of yes and no, on and off.

Shannon's information theory has been extended to account for quantum theory and the resulting framework, quantum information theory, has already shown a number of advantages. The greater power of quantum information theory is manifested in more secure cryptographic protocols, a completely new order of computing, quantum teleportation, and a number of other applications that were simply not possible according to Shannon's view. However, as quantum information theory is ultimately an extension of Shannon's information theory, under the right conditions the quantum information theory reduces to Shannon's. We also saw some tantalizing indications that biological systems may utilize quantum information to make some processes, such as photosynthesis, more efficient than anything possible according to our classical understanding of information.

The main aim of this book is how to understand reality in terms of information. In this respect it is appropriate to view the whole Universe as a quantum computer, given that this is our most accurate description. Then we estimated the total power of the Universe, a memory of 10 to the power of 100 bits and approximately 10 to the power of 90 bits processed per second. This estimate was possible by dividing the Universe into smaller and smaller units and then making use of the fact that the information content of each of these units is then proportional to its surface area.

So where does the information actually come from? When two people communicate one of them generates the information for the other. Any information in economic or social contexts comes likewise from human interactions. The information in human inter-action, i.e. biological systems, comes from the molecular properties of DNA. The behaviour of molecules is ultimately governed by the laws of quantum physics. In this way we can reduce any information that comprises reality down to quantum information. However, we are then still left with the question of where the quantum informa-tion comes from.

We now come back to the idea that the whole Universe is digital and we need to decode it in order that we can compress all the infor-mation into our laws; laws from which our reality then emerges. The fact that reality is somehow encoded into these laws is by no means novel. The Ancient Greeks, as we saw with Archimedes, understood the Universe that way, as did one of the first 'proper' scientists, Galileo Galilei.

Here is a quote from Galileo clearly expressing the view that the truths in the Universe are encoded into mathematics: 'Philosophy is written in this grand book – I mean the Universe – which stands continually open to our gaze, but it cannot be understood unless one first learns to comprehend the language in which it is written. It is written in the language of mathematics, and its characters are triangles, circles, and other geometric figures, without which it is humanly impossible to understand a single word of it; without these, one is wandering about in a dark labyrinth.'

But we want to go beyond Galileo's sentiment in two key respects. First, we want to use information instead of geometric characters. Second, we want to explain how the information in the Universe arises. Once the information is decoded and compressed into appro-priately defined laws, we can then understand our reality according to the information encoded in these laws. The laws themselves must be an integral part of this evolving picture. Otherwise we are stuck in an infinite regression. The Universe can therefore be seen as an infor-mation processor, in other words a gigantic quantum computer.

This view that the Universe is a computer is also not novel. Konrad Zuse, a famous Polish mathematician who pioneered many cryptographic techniques used during World War II, was the first to view the Universe as a computer. He was followed by a number of other researchers, most notably by the Americans Ed Fredkin and Tom Toffoli, who in the 1970s wrote a number of papers on this very topic. Fredkin is still seen as the leading champion of the digital model of the Universe and its internal workings. The problem, however, is that all these models assume that the Universe is a classical computer. By now, however, we know that the Universe should be understood as a quantum computer.

Our reality evolves because every once in a while we find that we need to edit part of the program that describes reality. We may find that this piece of the program, based on a certain model, is refuted (the underlying model is found to be inaccurate), and hence the program needs to be updated. Refuting a model and changing a part of the program is, as we saw, crucial to changing reality itself because refutations carry much more information than simply confirming a model.

These refutations are manifested as 'no-go' principles. Physics is littered with them. The Second Law of thermodynamics, which we saw was one of the most general laws of physics, is phrased to prohibit any transfer of heat from a cold to a hot body without any other effect. So, the Second Law would say that while we do not stipulate what physical processes can do, we certainly do know what they cannot do. Whilst we know the 'known knowns' and 'known unknowns' we do not know the 'unknown unknowns'. And this is very powerful, because it is extremely general. The same is true for the theory of relativity, as relativity tells us that you cannot travel faster than the speed of light.

When it comes to quantum mechanics, this 'no-go' way of speaking stretches our imagination to its limits. When we said that quantum mechanically an object can exist in two different places at once, this state is very difficult to understand using our everyday intuition. In fact, if we use the negative way, we are forced to acknowledge that (in some sense) it is not true that 'the object is in two places at once' and it is also not true that the object is 'not in two places at once'. So the statements that 'an object is in two places at once' as well as its opposite 'an object is not in two places at once' are both untrue. How can that be? It seems logically impossible that a statement and its negation are both incorrect. While to some this may be a contradiction, to Bohr this pointed to a deeper wisdom. He is reported to have said: 'A shallow truth is a statement whose opposite is false; a deep truth is a statement whose opposite is also a deep truth'.

However, we do not really need to change the laws of ordinary logic to resolve the quantum conundrum. There is no contradiction here as the two statements refer to two different experimental procedures. When we say that it is not true that the particle is in two places at the same time, this really refers to our detection procedure. When we measure the position of the particle, the particle is always really recorded in one, or the other place, but never in both. This confirms that we indeed have a particle. However, when we do not measure at all, but instead interact with the object in a way that does not record its location, then the object behaves as if it was in both places at the same time. So, two different handlings of the object will reveal two different scenarios for its behaviour. And there is nothing really contradictory about this fact.

However, the emerging reality does depend on what question we ask. We can force objects to adopt different characteristics depending on which particular characteristic we measure. All quantum information is ultimately context dependent. Einstein really did not like this view of reality, which somehow tells you that reality is created through your observations and is therefore not independent of us.

Interestingly enough, there is a very close theological position to the general Popperian philosophy of science and this position is known as the *Via Negativa* (or the negative way). It was apparently held originally by the Cappadocian Fathers of the fourth century, who based their whole world-view on questions which cannot be answered. For example, they proclaimed that, while they believed in God, they did not believe that God exists. This may appear to be a great contradiction, but it really is not.

As a matter of fact, the negative way was also well known in the East. In Hinduism, the idea of approaching god in terms of *Neti*, Sanskrit for 'not this', is very well established and documented from several ancient traditions, including Advaita Vedanta (which also specifies the Universe as single and inseparable, Brahman, whose features can only be grasped in the negative way).

The Cappadocian Fathers believed that one should describe the nature of God by focusing on what God is *not* rather than on what God *is*. The basic premise of this 'negative' (also called apophatic, for Greek 'what is not') theology is that God is so far beyond human understanding and experience that the only hope we have of getting close to the nature of God is to list all his negative features. And therefore we cannot say that God exists, because existence is a human notion and as such it may not apply to God.

This list of what God is not, compiled by the Cappadocian Fathers, is certainly rather reminiscent of the laws of physics and the general spirit of science. Physicists cannot really tell you what the Universe is or how exactly it behaves (or to be sure, we cannot tell what the ultimate description will be). But we can certainly tell you what it is not. We know that the Earth was not created 4000 years ago. It was much earlier, but we don't quite know exactly when. We don't exactly know how the Earth was created, but we do know that it did not arise by a giant turtle lifting its back above the surface of a cosmic ocean (or that the Earth was not formed before the Sun).

We also believe that the laws of physics are not different on Earth to anywhere else in the Universe, though we don't quite know what the ultimate laws of physics are (maybe quantum, maybe beyond quantum). Science, likewise, cannot really tell us fully about the ultimate origin of everything. Science is constructed more in a way that it tells us what the Universe is 'not like' rather than what it is like. For example, science tells us that we should not think of the Earth as in the centre of the Universe. Or we should not think of humans as the central point or purpose of evolution. It does not tell us exactly how we should think of humans, but we definitely should not think of them as fundamentally any different to apes, for example.

The Cappadocian Fathers reach the ultimate knowledge of God in the same way that we reach the ultimate understanding of reality. They do it via saying what God is not, whereas we rely on the scientific method of conjectures and refutations which tells us what reality is not. Although the negative way of speaking is in religion frequently (and I think mistakenly) viewed as irrational mysticism, we see that, in fact, it has a very rational basis paralleled in the scientific method.

Through this negative way of describing reality, separating that which is not true from everything else, we compress reality into a set of laws. These laws are then used as correct until proven otherwise. The laws of physics are the compression of reality which, when run on a universal quantum computer, produce reality. But the compressed laws of physics still need this universal quantum computer to produce reality. Since we are trying to explain the origin of everything, where does that computer come from?

But it's even more dramatic than that. We have the laws of Nature, which when run on a computer produce reality. So the laws of Nature need a pre-existing computer to be effective. On the other hand, the functioning of the computer itself needs to be described by some kind of laws. So what came first, the quantum computer or the physical laws? We seem to have a kind of 'chicken and egg' problem.

However, can the laws and the computer be created at the same time out of nothing? This possibility seems to be very difficult to imagine, but there have been many attempts at it throughout history. One way of doing so, that at least to me does not seem to be very elegant, or even scientifically valid, invokes what is known as the anthropic principle.

The anthropic principle states that the laws of the Universe are the way they are, because if they were different, we would not be here to talk about them. This argument might sound circular to you, but it is not. Any circular argument is, by default, logically correct (although trivially so, because they postulate what they later try to prove; think about the statement 'I like Jennifer Lopez, because she is my favourite person.') but the anthropic principle may just happen to be plain wrong. We simply do not know if some other laws would also not lead to beings, like us, who can do science and discover laws of Nature. A more modern improvement on the anthropic principle has been promoted by Sir Martin Rees, the British Astronomer Royal and the current president of the Royal Society. In Rees' version, on which he once jokingly bet his dog's life, all possible Universes exist and we only exist in the subset of those that have the right conditions. Let's hope for his dog's sake that he is correct.

Another possible answer, as we have seen, is that someone created the laws and the computer to start with. Traditionally, God has been assumed to be the original information creator. Alas, then we have to explain the origin of God, which is equally difficult.

But just take computers, for instance. In our computers we can design different worlds. Every computer game is, in fact, a simulation of a world with possibly very different rules to our own. This is what makes games very exciting and possibly also very difficult. As games and computers become more and more sophisticated the graphics get better and it becomes more difficult to tell a simulation from the real thing (enter Keanu Reeves, *The Matrix*).

However we still can't get away from the reliance on God. God in this case is the computer programmer who programmed the software that gives us our Universe. The reason why this answer sits so uneasily with scientists is – as explained – that, although we have an arguably prettier picture, it really only just displaces the question of the origin of information. So, if you tell me that we are someone's simulation, then who created that someone and told them to simulate us?

All answers of this type quickly lead to what is called an infinite regression, which we encountered earlier. For every creator, we seem to need to invent another creator that created it. God created our Universe, and another god created the first god and so on ad infinitum…This is no real solution, and cannot form the basis for answers to questions such as 'How come reality exists?'.

A similar problem of infinite regression led to von Neumman's theory of replication. There he faced the problem that one being has to contain a copy of the next generation which has to contain a copy of the next generation, and so on. It is clear that this kind of logic cannot be sustained in Nature because it requires you to store every future copy in the current version (i.e. an infinite amount of information in a limited amount of space). We simply have to look for some other possible explanation.

Throughout the ages it is interesting to note how the image of a god has changed in line with human knowledge. Prehistoric humans saw a different god in every element of life, and each god had to create that element and thereafter be responsible for it. With the Ancient Greeks, the notion of god was still polytheistic but their gods were smaller in number and correlated to more abstract notions such as love, war, peace, and happiness. There were also several religions that were pursuing a monotheistic view of god, for example Judaism, Christianity, and Islam. In the East we likewise encounter various versions of the principle of cosmic unity (though, in contrast, popular religion was presented in polytheistic terms).

The monotheistic view has survived for over two millennia but God's profession has changed somewhat. To Johannes Kepler, in the sixteenth century, God was a geometer, while for Newton, a century later, God was a physicist who, after creating the laws of physics together with the Universe, sat back and watched reality evolve. In Newton's world, the laws of physics are entirely deterministic and everything happens in an ordered fashion. In the latest incarnation of this story, we are talking about God as a computer scientist, sitting down and programming the Universe.

Although none of these roles that God has played can answer the question of the origin of information, there seems to be a distinct trend to God's role – namely, God seems to be less and less involved. Back in Ancient times, God had to create every little thing in the Universe and was responsible for its subsequent functioning – so you could say he had his work cut out for him. Then with Newton, on the other hand, God only had to create the laws of physics – once God had done this, he could just sit back and relax. It is therefore natural to ask if it's possible to reach a point where creation is so effortless that perhaps we don't even need a Creator.

There is a nice parallel in mathematics to what we are trying to argue for here, which is simply a version of the 'creation out of nothing'. A fascinating method of creating natural numbers out of empty sets was devised by von Neumann in the 1920s. Here is what he imagined. A set is a collection of things (just like the Universe). An empty set is a collection that contains nothing at all – you can think of this as zero information. Von Neumann proposed that all numbers could be bootstrapped out of the empty set by the operations of the mind.

While this may seems a little odd initially, there is a beautiful logic to it. The mind observes the empty set. It is not difficult to imagine this empty set also containing an empty set within itself. But hold on, now we have an empty set containing an empty set, so does this means that the original set has an element (albeit the element is an empty set)? Yes, the mind has thus generated the number one by producing the set containing the empty set. If we then consider that the empty set contained within the empty set yet contains its own empty set, then the mind has thus generated the number two out of emptiness. It is a set containing the set with nothing in it, and the set with the set with nothing in it (I hope your head is not spinning by now!). And so it continues ad infinitum. In this way, the mind creates all natural numbers, but literally out of nothing – starting from only an empty set. Starting with no information can, using von Neumann's logic, surprisingly lead to a great deal of information. All natural numbers (and there are infinitely many of them) can therefore be created out of an entirely empty set. In other words, we seem to have created an infinite amount of information from zero information.

Note that in von Neumann's beautiful vision, every subsequent creation depends on the previous one. There is a long chain of inter-dependent (correlated) creations. Each time the mind makes a decision to view the empty set in a different way, a new number appears. Correlations are thus very fundamental to our description and understanding of von Neumann's logic. However outside von Neumann's logic they are also crucial in the real world and manifest themselves through mutual information. It is, in fact, tempting to say that things and events have no meaning in themselves, but that only the shared (mutual) information between them is 'real'. All properties of physical objects, including their very existence, are only encoded in the relationships between them and hence in the information they share with other physical objects. This is not a particularly new view; this philosophy is already well developed and goes under the general name of 'relationalism'.

Eastern religion and philosophy have a strong core of relational thinking. In Buddhism, in particular, there is the notion of 'empti-ness' that is akin to von Neumann's empty set. What emptiness means in Buddhism is that 'things' do not exist in themselves, but are only possible in relation to other 'things'. For example, think of a chair. What is it really? There is a whole branch of philosophy, known as ontology, devoted to the questions such as 'What does it mean to be?', or 'What exists and in what sense is it real?'. I apologize in

advance to any ontologists out there and request their forgiveness as
I continue to use the minimal amount of technical precision to make
what I feel are the salient points needed for subsequent discussion.

Let us all imagine that a chair is just a collection of individual
parts, such as armrests, the seat, and so on. Well, it could be, but all
of these are just labels. Armrests and seats do not really exist independently of the context, e.g. you cannot have an armrest without
the concept of a 'chair' (let's assume only chairs have armrests) or
the concept of an 'arm'.

In search for the essence of 'chairness', that which defines a chair
independently of anything else, can we not just say that a chair is a
collection of atoms in the shape of a chair? But ultimately an atom is
also a label for a system containing some positively and some negatively charged particles as well as some neutral particles. And these
have all been labelled by us (an electron, a proton, and a neutron). If
you ask what an electron is, the answer would be a small negatively
charged particle, but all this is just a big label that tells us how this
particle behaves in various experiments (such as, it repels some
particles and attracts others). It is ultimately a 'label' to describe the
various sorts of behaviour that electrons exhibit when we try to
interact with them and manipulate them. Without this label we
would have to call an electron something like 'You know that particle
that does X when we test Y and it does P when we look at it in Q, and
so on'. In this way we can see that labels are awfully convenient and
efficient. But Buddhism tells us that we should not confuse the label
with the object. More importantly, just because we have a label for
something, it does not mean that this something is real.

Quantum physics is indeed very much in agreement with
Buddistic emptiness. The famous British astronomer Arthur
Eddington put it this way: 'The term "particle" survives in modern
physics but very little of its classical meaning remains. A particle
can now best be defined as the *conceptual carrier of a set of variates*...It
is also conceived as the *occupant of a state* defined by the same set of

variates… It might seem desirable to distinguish the "mathematical fictions" from "actual particles"; but it is difficult to find any logical basis for such a distinction. *Discovering* a particle means observing certain effects which are accepted as proof of its existence.'

Eddington claims here that a particle is just a set of labels that we use to describe outcomes of our measurements. And that's it. It all boils down to a relation between our measurements and our labels! The complexity that we see around us in this world (and this complexity we believe to be growing with time, as far as life is concerned at least) is just due to the growing interconnectedness.

In this way, can we now analyse how we encode reality? By doing so, we will never arrive at 'the thing in itself' by any kind of means. Everything that exists, exists by convention and labelling and is therefore dependent on other things. So, Buddhists would say that their highest goal – realizing emptiness – simply means that we realize how inter-related things fundamentally are. Exactly the same is true in other Eastern religions. Less well known in the West is Advaita Vedanta – a Hindu philosophy that emphasizes the total oneness of the Universe. In this view our perceptions of separate entities is just an illusion – *Maya*. Even the Universe as a whole only exists by labelling and not by itself. Our reality is 'that which is the sum total of all the observations and facts humanity has gathered so far'!

We have reached a point where any particle of matter (such as an atom) and energy (such as a photon) in the Universe is defined with respect to an intricate procedure that is used to detect it. If the detector makes a click (like a Geiger counter) the particle is detected. The click itself creates one extra bit of information comprising reality. The crucial point is that the particle does not exist independently of the detector.

But what exactly constitutes a click for any detector such as a Geiger counter? A click is the positive result of any experimental procedure capable of detecting the presence of a particle. This is done by generating a specific interaction between the experimental apparatus and the spatial region in which we are searching for the particle. The interaction needs to be carefully engineered – some interactions simply will not be able to serve this purpose, i.e. they will not give us the relevant bit of information. Going back to the photon example, a beam-splitter will not give us any information as to the existence of the photon. Nothing in the beam-splitter retains any record of the presence or absence of the photon that might have gone through. In other words, beam-splitters cannot produce clicks. If we want a click we have to use something instead of or in addition to beam-splitters, such as photomultipliers. Photomultipliers are designed so that the presence of a photon knocks out one electron, whose motion triggers an electrical current whose presence is amplified to the macroscopic level. This results in a click that we can hear, or any other effect that we can register with our senses.

We can push this a bit further, however. We can ask if the particle is the cause for the detector to click. The answer is no. The reason is that in quantum physics, as we have argued, particles exist and don't exist at the same time. Here I don't just mean that they exist in different places, I mean that even in one place a particle can exist and not exist simultaneously. This too is a direct consequence of quantum indeterminacy. What would this mean in the beam-splitter example discussed earlier? It means that a photon simultaneously enters and does not enter it and this implies that the only time it will be detected at the output is when it does exist. Whether the detector clicks or not is a genuinely random event that cannot be predicted by any means, in the same way that we cannot predict the photon's reflection at the beam-splitter. This implies that we should not say that an existing particle causes the click just like we cannot say that photon's reflection caused a click (since we know that it also passes through). The click has no cause at all and therefore we have no underlying particles.

And since there are no underlying particles in reality, there are no things in the Universe that are made up of particles existing without the intricate procedures to detect them. Detection events are genuinely random and the emerging reality is seen in the correlations, expressed as laws of physics, between the events, which are bits of information. If the link between the information compression and randomness is as Kolmogorov and Chaitin thought, then this conclusion is likely to hold no matter what underlying theory of Nature we discover in the future.

Reality is made up of quantum bits, each arising from a causeless click. A click entirely without a cause has the novel property that it introduces a discontinuity in time. Once an event is recorded, it is solidified forever in the Universe. It becomes an element of what we call the past. However, before the event occurs, we have an uncertainty as to when and if it will happen. All possibilities are then present at the same time and the game is completely open. The occurrence of the event then belongs to something we call the future. Fundamental randomness at the core of reality therefore allows us to have a distinction between the solidified, unchanging, past and a fluid, dynamic, future.

The distinction between the past and future separated by a discontinuity due to a measurement is always relative to the observer who recorded the measurement click. Someone who is able to control the observer and his interactions with the environment, would, according to our current understanding of quantum mechanics, be able to reverse the detection and thereby erase the observer's past. There is no contradiction here, it is purely an interplay between local information (that of the observer) and global information (that of the person who reverses the observation by manipulating the observer and environment).

An amazing issue to note is this. The above meditation in realizing emptiness is a very similar exercise to von Neumann's number creation out of empty sets; it just goes in the exact opposite direction. Von Neumann went from an empty set into an infinite set of real numbers and here we started with macroscopic objects and deconstructed them to find that actually there is nothing behind them, they are based on randomness, on no prior information.

This is the darkness of reality! Anything that exists in this Universe, anything to which you can attribute any kind of reality, only exists by virtue of the mutual information it shares with other objects in the Universe. Underneath this, nothing else exists, nothing else has any underlying reality and hence there is no infinite regression. It just has to be this way, as otherwise we are asking a finite Universe to contain an infinite amount of information – and this is clearly not possible!

Following this logic, it's more accurate to think of the evolution of the Universe as starting with all potential realities from which one reality simply emerges. From this initial state, which contains all possible subsequent futures, the first event occurs without any cause (i.e. a random event) and this gives us our first bit of information. So from all possible futures, now we have a smaller number of futures simply because the first event has occurred in a specific way and all subsequent events will have this event as their past. In this way mutual information is established between bits.

Here we can draw an analogy with sculpting. A sculptor starts with a block of stone out of which he intends to make a sculpture. In some sense, we can say that the initial untouched block contains all the possible sculptures to be made. This is a bit like our initial Universe where all possible realities exist at the same time, but are not actualized. The sculptor then makes the first move, and chisels a piece away from the block. This first cut by the artist breaks the symmetry and reduces the information contained in the initial block. We now no longer have all possible sculptures available to us as some of them, which required the piece that was chiselled away, can now no longer be realized.

For example, think of Michelangelo standing in front of a six metre block of stone, just about to begin work on his sculpture of David. David now stands proudly at five metres tall in the Galleria dell'Accademia in Florence, the masterpiece of a Renaissance genius. Imagine if the first move of this genius was a slip, where he cuts horizontally, so that he now has two smaller blocks each of height three metres. So he can still make a statue but would no longer be the same David. Imagine now that he makes another mistake, this would clearly impact what could or couldn't be created from the remaining block of stone. His possibilities are therefore reduced.

And so it continues, with every next move that the sculptor makes the number of possible futures decrease. Once the sculpture is finished, one possibility is crystallized. Even then, there are more things that could be done to change the sculpture, so we never really arrive at something final. Whenever we think we have something final, the sculptor can always make another cut. Of course, what happens when there are no cuts left to be made? Is this a realistic scenario? From what we have argued, this will never happen in the Universe, as a change of perspective generates new ideas and new information as to what the shape of reality could be. Through successively smaller and smaller alterations the sculptor will always be able to make another cut to whatever is left.

This very way of thinking about the Universe completely and faithfully embodies the spirit of how science operates. We gather information about the Universe by observing different things and these observations then go on to shape our reality. In this way reality emerges around us in a definite and concrete manner. However, since the information we gain from the Universe is defined through an observer, the question is how do we define this observer – do we have a universal observer whose observations we can trust beyond any doubt? Well, apart from bringing in the concept of a supernatural being (which is always a bit of a cop-out), every other observer seems as good as any other to define his or her reality.

In Chapter 2 we found ourselves defining reality through Calvino's card game. In this card game each player represents an observer, and in turn each observer represents a different aspect of reality (economics, physics, biology, social science, computer science, and philosophy). Each observer communicates what he has experienced through a sequence of playing cards. Physics would be telling us about physical laws, such as, if we drop an apple it will fall to the ground, or if we heat water beyond a certain temperature it will become steam. In the same way economics, biology, and all the other observers will also have their own story to tell.

Each observer at the table, as well as telling their own story, will listen to every other player's story. In this way reality emerges through the sharing of information between the players. For example, to date, the cards that have been revealed by physics indicate that the fastest speed of travel is the speed of light. This is not to say that in the future physics may not show us another card which then tells us that under certain conditions we could travel faster than the speed of light. Whilst the players are consistent in their stories, our interpretation of what we are observing evolves to a better and better approximation the longer we are given to observe. Just as in an everyday conversation, only hearing half the story inevitably only gives you a fraction of the information – and you may even get the wrong message entirely. Unfortunately, to wait for this game to be finished before making sense of its information content would require us to wait for the age of the Universe – so instead we are continually guessing at defining reality as best we can.

Each of the players communicates their story through the cards. These cards are assumed to be predefined (like a language) and enable the players to communicate their story. From our previous discussions can we say anything about where these cards come from? Actually we can. We have seen previously in this chapter that although information does come in discrete units (such as cards or bits of information), actually these discrete units are based on a fundamental level of randomness. So if the cards themselves have some element of randomness (i.e. sometimes a card could represent 'force' and sometimes it could represent 'peace') then how is it that anyone is able to tell a consistent story with these cards? Surely it would be impossible to tell a story with cards which are not well defined – it just doesn't seem logical! It is counterintuitive that although we seem to perceive a well-defined reality around us, quantum physics suggests that there is no underlying single reality in the Universe independent of us – and that our reality is actually only defined if and when we observe it.

For example, when a particle of light, a photon, encounters a piece of glass like your bedroom window, two outcomes can occur. One outcome is that the photon can be reflected, and the other one is that it can pass through the window. Quantum physics tells us that if we observe the photon we will never be able to predict the outcome in advance, this process is completely random. But what happens if we do not observe the photon? Then quantum physics suggests that the photon takes both alternatives: it both goes through the glass and is also reflected – it exists in two places simultaneously, i.e. there are two distinct realities!

But we only seem to see one reality around us, you never see the same person existing in two different places at the same time. So how does the act of observation allow one reality to emerge out of two or more realities? Quantum physics seems to imply that reality somehow emerges through interactions between observers and the observed. This is reminiscent of a magician's trick, where the main point is to make a card appear from a pack of cards within which it didn't exist. To make this point clearer, let me convey the same message through a simple game.

Suppose you have four players, each of whom is given four cards at the beginning of the game. The goal of the game is for a player to obtain four of the same cards (four aces, four tens, etc.) by exchanging cards with other players. The first person to do this is the winner. There are, however, a couple of rules. The first rule is that you can only ask for a card if you already have at least one of those cards. So you can only ask someone for an ace, if you are holding at least one ace. Now if the person says no, that means they do not have an ace and the game passes to the next player. If they do have the card, they must give it to you and then you have the option of asking the same player again or another player. When you ask for an ace, all other players immediately know that you have at least one ace. The next player in turn then knows what to get from you. Of course this is a very simple game and really doesn't need more than a couple of rounds to find a winner.

The amazing thing is that we don't really need cards to play the game. This is where it gets interesting. The whole game can be played just inside the players' heads where each player imagines four completely arbitrary cards, which is non-standard, in that there is no limit on what image can be on any card or how many of these images exist. For example, one person could imagine three elephant cards and a crocodile card, whilst someone else could imagine two aces and two apple cards. As long as we add the requirement that you can't start with four of the same cards then we know that you must be forced to ask at least one question. Surprisingly whilst there seems to be an infinite number of combinations, this is actually not a problem and we can always find a winner. The saving grace is that the players cannot change any choices that would affect the consistency of the game so far, although they can change their cards throughout the game. For example, if you've asked someone else for a card, then you must at least possess a card of that type. And if someone has asked for a particular card then they must give up that card if it has been asked for by someone else. This requirement of consistency and the ability to change your cards to win the game is what quickly narrows down the multitude of different possibilities as the game proceeds. Both answering and asking a question affects both the cards that you hold and the cards the other players hold. In this way this imagined version without physical cards or limits reduces to the fixed game and we ultimately end up with a definite winner.

This constant questioning is analogous to experimentation in physics, where we start by observing an infinite number of possibilities; however through interaction with the system and by modifying our experiments in line with the information available, this reduces to a smaller set of possible outcomes and then to an eventual winner, a single reality. Experiments too have to be consistent with the rules of the game, the laws of physics. Reality is therefore created by experiments in the same way that cards become created in this imaginary card game. Through this analogy, I hope the reader gets a feel for the bizarreness of quantum mechanics.

I said that information in the Universe is indeed very much like Calvino had imagined. It is discrete (and we saw why this is advantageous when we talked about DNA and life), it is context dependent and it is finite. The crucial aspect missing in Calvino's card game is that in reality, there are no cards. Nature does not give us cards to start with; this is a special no-go theorem of quantum physics excluding the so-called hidden variables. We have to create the cards ourselves through observations.

The proper analogy between cards and information in the Universe is a combination of Calvino's card game with the one we described earlier. Imagine that each player has a different set of cards, and each card has no predefined meaning. This means that just like the sculptor, who when he starts with the block of stone can carve any shape, we have the initial conditions to define any reality that we care to, because at the beginning of the game all possible realities can potentially exist.

As you question other people to gather your cards, your life story unfolds in an unpredictable way, a way that depends on what cards other people tell you they possess. This process is not fixed by anything other than the consistency of the stories they have already told. For example, if physics told us that it dropped an apple which fell to the ground, then it cannot later change its story that the apple did not fall to the ground – the event has been defined and communicated and now it exists on record and cannot be changed. If we later discover that under some circumstances the apple does not fall to the ground, this does not contradict the player's story, it just adds a new insight – in that under certain conditions it may not fall.

The main feature of the 'card game without cards' is precisely the fact that cards emerge out of nowhere. We start with no information at all (or with infinite information, as is more appropriate, since all possibilities are open). Anything is possible as far as card arrangements and signs are concerned. Then we start questioning and out emerges a definite order. Questions, because they are subject to a (very small) set of rules, reveal a certain type of reality that was not there before (or without) the questions. In a typical card game all the cards are fixed when dealt, but this is clearly not so in our game.

Touching upon that sensitive issue of having cards without any cards existing in the first place, the British chemist Peter Atkins offers an explanation: 'In the beginning there was nothing. Absolute void, not merely empty space. There was no space; nor was there time, for this was before time. The Universe was without form and void. By chance there was a fluctuation and a set of points, emerging from nothing and taking their existence from the pattern they formed, defined a time.' Then the space (and therefore cards) gets created in a similar way and the rest is history. This version of events looks appealing, but the trouble is that the initial fluctuation that leads to everything is difficult to quantify without any prior theory, i.e. without the rules of the card game. In order to define the size and probability of a fluctuation, we usually need more information, like, for example, knowledge of quantum theory, one of the key rules of the game.

Just as we need the laws of physics to describe events, events themselves need the laws of physics to happen. So, which of the two came first? If we imagine that the laws of physics came first and they then dictate how events unfold, are we really being consistent? Laws of physics become laws because there are events which consistently produce outcomes in accordance with those laws. Events are therefore the material onto which the laws are written. In order for the laws of physics to come first, this would mean that there were no prior events in accordance with this law and therefore the question remains whether this is a law at all, though as we have seen, mutual information can arise out of no overall information and events themselves can happen without any prior rule.

Leibniz's logic was that the simplest possible state of the Universe is the one that contains nothing, so the fact that we see something is for him the best proof for the existence of God. However, in our picture, having nothing at the beginning corresponds to no information. In Shannon's theory this would mean zero entropy of the whole Universe. Any subsequent information gain is not necessarily proof for the existence of God because, as we have seen, mutual information can be ultimately generated locally even though the information overall remains zero.

We can construct our whole reality in this way by looking at it in terms of two distinct but inter-related arrows of knowledge. We have the spontaneous creation of mutual information in the Universe as events unfold, without any prior cause. This kicks off the interplay between the two arrows. On the one hand, through our observations and a series of conjectures and refutations, we compress the information in the Universe into a set of natural laws. These laws are the shortest programs to represent all our observations. On the other hand, we run these programs to generate our picture of reality. It is this picture that then tells us what is, and isn't, possible to accomplish, in other words, what our limitations are.

The Universe starts empty but potentially with a huge amount of information. The key event that gives the Universe some direction is the first act of 'symmetry breaking', the first cut of the sculptor. This act, which we consider as completely random, i.e. without any prior cause, just decides on why one tiny aspect in the Universe is one way rather than another. This first event sets in motion a chain reaction in which, once one rule has been decided, the rest of the Universe needs to proceed in a consistent manner. Just like in Calvino's card game, the next piece of the story must be consistent with the previous.

This is where the first arrow of knowledge begins. We compress the spontaneous, yet consistent information in the Universe, into a set of natural laws that continuously evolve as we test and discard the erroneous ones. Just as man evolved through a compression of biological information (a series of optimizations for the changing environment), our understanding of the Universe (our reality) has also evolved as we better synthesize and compress the information that we are presented with into more and more accurate laws of Nature. This is how the laws of Nature emerge, and these are the physical, biological, and social principles that our knowledge is based on.

The second arrow of knowledge is the flip-side to the first arrow. Once we have the laws of Nature, we explore their meaning in order to define our reality, in terms of what is and isn't possible within it. It is a necessary truth that whatever our reality, it is based exclusively on our understanding of these laws. For example, if we have no knowledge of natural selection, all of the species look independently created and without any obvious connection. Of course this is all dynamic in that when we find an event that doesn't fit our description of reality, then we go back and change the laws, so that the subsequently generated reality also explains this event.

The basis for these two arrows is the darkness of reality, a void from which they were created and within which they operate. Following the first arrow, we ultimately arrive at nothing (ultimately there is no reality, no law without law). The second arrow then lifts us from this nothingness and generates a picture of reality as an interconnected whole.

So our two arrows seem to point in opposite directions to one another. The first compresses the information available into succinct knowledge and the second decompresses the resulting laws into a colourful picture of reality. In this sense our whole reality is encoded into the set of natural laws. We already said that there was an overall direction for information flow in the Universe, i.e. that entropy (disorder) in the Universe can only increase. This gives us a well-defined directionality to the Universe, commonly known as the 'arrow of time'. So how do our two arrows of knowledge stand in relation to the arrow of time?

The first arrow of knowledge clearly acts like a Maxwell's demon. It constantly combats the arrow of time and tirelessly compresses disorder into something more meaningful. It connects seemingly random and causeless events into a string of mutually inter-related facts. The second arrow of knowledge, however, acts in the opposite direction of increasing the disorder. By changing our view of reality it instructs us that there are more actions we can take within the new reality than we could with the previous, more limited view.

Within us, within all objects in the Universe, lie these two opposing tendencies. So, is this a constant struggle between new information and hence disorder being created in the Universe, and our efforts to order this into a small set of rules? If so, is this a losing battle? How can we ever compete with the Universe?

Key points

- Scientific knowledge proceeds via a dialogue with Nature. We ask 'yes-no' questions through our observations of various phenomena.
- Information in this way is created out of no information. By taking a stab in the dark we set a marker which we can then use to refine our understanding by asking such 'yes-no' questions.

- This inductive method, which is the basis of constructing physical theories, I called the darkness of physics. Stating that something is not is the key to building better and better models of the world. Physical laws are usually more fundamental the more they rule out: 'no such process exists, where so and so would happen' is a typical way of formulating it.

- There are many parallels in religion which use the Nagative Way to reach the ultimate truth. Two prominent examples are the Cappadocian Fathers in early Christianity and Advaita Vedanta in Hinduism.

- The whole of our reality emerges by first using the conjectures and refutations to compress observations and then from this compression we deduce what is and isn't possible.

EPILOGUE

This book has argued that everything in our reality is made up of information. From the evolution of life to the dynamics of social ordering to the functioning of quantum computers, they can all be understood in terms of bits of information. We saw that in order to capture all the latest elements of reality we needed to extend Shannon's original notion of information, and upgrade his notion from bits to quantum bits, or qubits. Qubits incorporate the fact that in quantum theory outcomes to our measurements are intrinsically random.

But where do these qubits come from? Quantum theory allows us to answer this question; but the answer is not quite what we expected. It suggests that these qubits come from nowhere! There is no prior information required in order for information to exist. Information can be created from emptiness. In presenting a solution to the sticky question of 'law without law' we find that information breaks the infinite chain of regression in which we always seem to need a more fundamental law to explain the current one. This feature of information, ultimately coming from our understanding of quantum theory, is what distinguishes information from any other concept that could potentially unify our view of reality, such as matter or energy. Information is, in fact, unique in this respect.

Viewing reality as information leads us to recognize two competing trends in its evolution. These trends, or let's call them arrows, work hand in hand, but point in opposite directions. The first arrow orders the world against the Second Law of thermodynamics and compresses all the spontaneously generated information in the Universe into a set of well-defined principles. The second arrow then generates our view of reality from these principles.

It is clear that the more efficient we are in compressing all the spontaneously generated information, the faster we can expand our reality of what is and isn't possible. But without the second arrow, without an elementary view of our reality, we cannot even begin to describe the Universe. We cannot access parts of the Universe that have no corresponding basis in our reality. After all, whatever is outside our reality is unknown to us. We don't yet know what we don't know!

But let's try to look beyond this, into the unknown. What if the second arrow, which generates our view of reality, somehow affects the first arrow – our compression of the information that the Universe gives us? It is not so surprising that this relationship has been the key in the evolution of our reality thus far. By exploring our reality we better understand how to look for and compress the information that the Universe produces. This in turn then affects our reality. Everything that we have understood, every piece of knowledge, has been acquired by feeding these two arrows into one another. Whether it is biological propagation of life, astrophysics, economics, or quantum mechanics, these are all a consequence of our constant re-evaluation of reality. So it's clear that not only does the second arrow depend on the first, it is natural that the first arrow also depends on the second.

But if indeed they are mutually dependent, where exactly does this leave us? Not anywhere clear cut I'm afraid. Neither arrow can exist on its own and is somehow predetermined by its complement. Once the initial symmetry is broken and we get information out of no information, the first and the second arrow play out their roles within a self-perpetuating cycle. We compress information to generate our laws of Nature, and then use these laws of Nature to generate more information, which then gets compressed back into upgraded laws of Nature.

The dynamics of the two arrows is driven by our desire to understand the Universe. As we drill deeper and deeper into our reality we expect to find a better understanding of the Universe. We believe that the Universe to some degree behaves independently of us and the Second Law tells us that the amount of information in the Universe is increasing. But what if with the second arrow, which generates our view of reality, we can affect parts of the Universe and create new information? In other words, through our existence could we affect the Universe within which we exist? This would make the information generated by us a part of the new information the Second Law talks about.

A scenario like this presents no conceptual problem within our picture. This new information can also be captured by the first arrow, as it fights, through conjectures and refutations, to incorporate any new information into the basic laws of Nature. However, could it be that there is no other information in the Universe than that generated by us as we create our own reality?

This leads us to a startling possibility. If indeed the randomness in the Universe, as demonstrated by quantum mechanics, is a consequence of our generation of reality then it is as if we create our own destiny. It is as if we exist within a simulation, where there is a program that is generating us and everything that we see around us. Think back to the movie *The Matrix*, where Keanu Reeves lives in a simulation until he is offered a way out, a way back into reality. If the randomness in the Universe is due to our own creation of reality, then there is no way out for us. This is because, in the end, we are creators of our own simulation. In such a scenario, Reeves would wake up in his reality only to find himself sitting at the desk programming his own simulation. This closed loop was echoed by John Wheeler who said: 'physics gives rise to observer-participancy; observer-participancy gives rise to information; information gives rise to physics.'

But whether reality is self-simulating (and hence there is no Universe required outside of it) is, by definition, something that we will never know. What we can say, following the logic presented in this book, is that outside of our reality there is no additional description of the Universe that we can understand, there is just emptiness. This means that there is no scope for the ultimate law or supernatural being – given that both of these would exist outside of our reality and in the darkness. Within our reality everything exists through an interconnected web of relationships and the building blocks of this web are bits of information. We process, synthesize, and observe this information in order to construct the reality around us. As information spontaneously emerges from the emptiness we take this into account to update our view of reality. The laws of Nature are information about information and outside of it there is just darkness. This is the gateway to understanding reality.

And I finish with a quote from the *Tao Te Ching*, which some 2500 years earlier, seems to have beaten me to the punch-line:

> The Tao that can be told is not the eternal Tao.
> The name that can be named is not the eternal name.
> The nameless is the beginning of heaven and earth.
> The named is the mother of the ten thousand things.
> Ever desireless, one can see the mystery.
> Ever desiring, one sees the manifestations.
> These two spring from the same source but differ in name; this appears as darkness.
> Darkness within darkness.
> The gate to all mystery.

NOTES

Chapters 1–2

E.J. Larson and L. Witham, Scientists are still keeping the faith, *Nature*, **386**, 435 (1997). This article presents some statistics on religion among scientists. Though one has to be careful with such statistics, as depending on the exact wording the responses can be quite different. For example, the questions 'Do you believe in God?', or 'Do you believe in a supernatural being?' or simply 'Are you religious?' could (and do) all lead to different statistics of answers.

I. Calvino, *Castle of Crossed Destinies* (Vintage Classics, 1997). A creative parable on life by one of the foremost Italian writers. Calvino's card game is used as the main metaphor in my book, in terms of how we gain knowledge and better understand our reality. Various writers have provided different metaphors for life, in terms of games that we play; however Calvino's card game for me is richer and more insightful.

W. Poundstone, *Recursive Universe* (William Morrow, 1984). One of the first popular books to argue for the digital view of the Universe in a very eloquent and general way. As far as I am aware, the first person to think of the Universe as a gigantic information processor was a Polish computer scientist, Konrad Zuse, whose mathematics was instrumental for the Allied code-breaking activities in World War II. Unfortunately though, he never wrote any accessible account of it. Other notable protagonists include Tommaso Toffoli and Edward Fredkin.

Chapter 3

S. Wrathmell, *Leeds* (Pevsner Architectural Guides, Yale University Press 2008). An excellent guide to the architectural and cultural heritage of Leeds, UK, my home town between 2004 and 2009.

J.R. Pierce, *Information, Signals, Noise* (Dover, 1973). Pierce has written the best accessible account of information theory I am aware of. This requires some basic knowledge of mathematics, but only very basic! I strongly encourage you to read it if you are interested in delving deeper into elements of the information theory presented in this book.

C.E. Shannon and W. Weaver, *The Mathematical Theory of Communication* (University of Illinois Press, 1948). The Bible of information theory. The book contains both Shannon's original paper as well as a commentary by Weaver.

E.C. Cherry, A history of the theory of information, *Proceedings of the Institute of Electrical Engineering*, **98**, 383 (1951). My account was really just geared towards explaining the established theory of information, beginning with Shannon. This work however explains in great detail the history of the basic ideas that led to this established theory. A shorter review, with somewhat different emphasis, is by J.R. Pierce, The early days of information theory, *IEEE Transactions on Information Theory*, **19**, 3 (1973).

Chapter 4

J. von Neumann, *Theory of Self-Reproducing Automata*, edited and compiled by Arthur W. Burks (University of Illinois Press, 1963). After writing pioneering treatises in economics, quantum physics, and mathematics, von Neumann then turned his attention to biologically inspired questions. This book contains the original exposition of von Neumann's ideas about replication. Like most of von Neumann's work, a good read for the more mathematically astute readers.

E. Schrödinger, *What is Life?* (Cambridge University Press, 1946). A beautifully written popularization of physics, with emphasis on implications for biology. It is still highly recommended, even though many ideas there have since been surpassed.

J. Monod, *Chance and Necessity* (Vintage, 1971). This book views life as consisting of reproductions of Maxwell's demons. Passionately argued, and beautifully explained, by a Nobel Laureate in biology.

Chapter 5

B. Russell, *Free Man's Worship* (Routledge, 1976). In the agnostic tradition of Thomas Henry Huxley, Russell explains what a free person should and should not accept in the light of scientific knowledge. This contains the quote of Russell on the faith scientists and philosophers have in the Second Law of thermodynamics.

P. Atkins, *Creation* (Oxford University Press, 1978). This book beautifully argues how the tendency to chaos captured by the Second Law is in fact the main driving force behind evolution. Far from contradicting it, disorder generates life – which could be viewed as an oasis in the sea of chaos. The book is also notable for trying to provide a physical creation of the Universe out of nothing. But, as I argue in my book, this picture lacks the crucial concept of information that permeates all phenomena. This is the book Richard Dawkins said was the best popular science book ever written!

T. Norretranders, *The User Illusion: Cutting Consciousness Down to Size* (Penguin Press Science, 1998). This book contains a detailed description of Maxwell's demon paradox and its consequences for computation. It is otherwise one of the best attempts so far to understand consciousness in terms of information theory. According to Norretranders our brain makes representations of reality as images in our head. One aspect of this reality is ourselves and the evolving image of ourselves is, briefly speaking, our mind. The 'user illusion' in the title refers to the fact that the computer also creates an illusion of itself in order for us to find it user-friendly. So we think of computers as desktops with files, folders, programs, routines, etc. However, all a computer does is simply crunching of zeros and ones. Nowhere inside any computers do folders exist or programs or files – this is just an interface for us. Our consciousness likewise offers us an interface of ourselves. This is all there is to mind, is what the book claims.

Chapter 6

R.J. Kelly, A new interpretation of information rate, *Bell Systems Technical Journal*, **35**, 916 (1956). This is the first application of Shannon's theory and it is to gambling. One of its striking features is that no 'error correction' is necessary in order to reach the maximum capacity (in this case the maximum financial gain).

E.O. Thorp, *The Mathematics of Gambling*. Notes based on Thorp's preparation and experience in the Las Vegas casinos. Written in a very simple language and accessible online to a wide audience.

K. Sigmund, *Games of Life* (Oxford University Press, 1993). If you were fascinated by the analogy of betting in a casino, with organisms betting against the environment to survive, then this is an excellent popular book outlining the role of game theory in biology in a much more general context.

Chapter 7

T. Harford, *The Logic of Life* (Little Brown, 2008). Contains a section discussing Schelling's basic ideas. The whole book is beautifully written and examines various general issues from the perspective of an economist.

M. Buchanan, *Nexus: Small Worlds and the Groundbreaking Science of Networks* (W.W. Norton, 2002). This is a good up-to-date popular account of modern mathematical methods in sociology. Buchanan is a journalist and this is a very accessible introduction. If you want something a little more detailed, then I would recommend A.-L. Barabási, *Linked: The New Science of Networks* (Plume, 2003). Barabási is a scientific researcher who has contributed a great deal to studying physical properties of general networks.

J.E. Stiglitz, *Globalization and Its Discontents* (W.W. Norton, 2003). Here is a book describing the socio-economic implications of globalization. Stiglitz, an economics Nobel Laureate, is careful to point out the pros and cons as well as to offer advice on how to direct the globalizing trend towards the greater good of everyone, rather than increase the gap between the haves and have-nots. He does not talk about any connections with information theory though.

T.L. Friedman, *The World is Flat* (Farrar, Straus and Giroux, 2005). A personal account of a journalist of what it means to have a very inter-connected world. 'Flatness' in the title refers exactly to the fact that everyone is connected to everyone else and all changes propagate at a very high speed. A personal account of interconnectedness in the world, again without any use of information theory.

Chapter 8

W. Heisenberg, *Physics and Philosophy* (George Allen and Unwin, 1959). An excellent account of the basic tenets of quantum mechanics and how it changed the whole classical philosophy. Written by one of its pioneers, it is a *tour de force*.

B. Clegg, *The God Effect* (St. Martin's Press, 2006). This book contains a very accessible introduction to the recent work on entanglement, both theoretical and experimental. Highly recommended to all those interested in an up-to-date understanding of quantum mechanics.

S. Singh, *The Code Book* (Fourth Estate, 2000). A popular book explaining cryptography within the historical setting. Many intriguing examples are presented. The author also reviews the basics of quantum cryptography.

Chapter 9

D. Deutsch, *The Fabric of Reality* (Allen Lane, The Penguin Press, 1997). A very creative account of our current understanding of reality through four pillars of knowledge: quantum physics, biology of the selfish gene, Popper's conjectures and refutations, and Turing's theory of universal computation.

H. Everett, *Relative State Interpretation of Quantum Mechanics* (Princeton University Press, 1973). The first application of Shannon's information theory to quantum mechanics. For Everett, measurement results are in fact stored in the correlations between the observers and the observed. The correlations are measured using Shannon's formula. The overall state of the Universe is presented as a huge combination of correlated states between different subsystems. What matters is the state of one relative to another, hence the title of Everett's thesis. This relational view of the Universe forms the basis for the view I offer in the last chapter of the book.

G.R. Fleming and G.D. Scholes, Physical chemistry: Quantum mechanics for plants, *Nature*, **431**, 256 (2004). A friendly, semi-popular one-page piece on the potential importance of quantum effects in biology. This is now a growing area of research.

Chapter 10

P. Watson, *Ideas* (Phoenix, 2005). A recent book arguing that there are three key ideas in the development of Western civilization. The scientific method, or method of conjectures and refutations, is one of them.

M. Schroeder, *Fractals, Chaos, Power Laws: Minutes from an Infinite Paradise* (W.H. Freeman, 1992). Schroeder is brilliant at conveying simple ideas behind randomness in an exciting way so that even the expert remains entertained. Highly recommended.

K. Popper, *Conjectures and Refutations* (Routledge, 2002). Popper is the most-liked philosopher by scientists because he clarified and defended the method by which scientists gain knowledge.

G. Chaitin, *Collection of Essays* (World Scientific, 2007). Essays mainly on the topic of viewing randomness in an information theoretic way. Closely related to the work of R. Solomonoff, A formal theory of inductive inference, *Information and Control*, **7**, 1 (1964).

V. Vedral, 50th Anniversary Issue of the *New Scientist* (18 Nov. 2006). My invited essay on determinism versus randomness from the physics perspective. Some parts of this chapter are based on this essay.

Chapter 11

Archimedes, *The Sand Reckoner* (a translation can be found on the web). An essay with a very visionary calculation undertaken by the Ancient Greek mathematician (Archimedes usually features with Gauss and Newton in the top three mathematical geniuses of all time). He offers to the then king of Syracuse the reasoning behind his estimate of the size of the Universe. He puts it in terms of the number of grains of sand (probably what was thought to be the smallest object at that time) that can fit within the Universe. It is intriguing to read how he enumerates large numbers. Remember that the Greeks didn't have the concept of zero. So he could not write 1000,000 to represent a million!

L. Smolin, *Three Roads to Quantum Gravity* (Basic Books, 2002). A good popular account of the Bekenstein bound and the relationship between entropy and area.

J. Barbour, *The End of Time* (Oxford University Press, 2001). This book argues that since all meaning lies in connections between events, time itself does not exist. That is, it does not exist over and above the correlations. Paraphrasing it in the spirit of the current book, time is just the amount of correlations between things in the Universe. The book is based on a well-known paper by D.N. Page and W.K. Wootters, *Physical Review D*, **27**, 2885 (1983).

Chapter 12

D. Turner, *The Darkness of God* (Cambridge University Press, 1995). This book explains the basics of medieval Christian mysticism. Contrary to what we perceive mysticism to be today, the medieval Christian variant was very rational and against the so-called 'mystical experiences' as necessary to reach God. The key was a consistent application of the Negative Way, a method they invented that has similarities with the scientific method.

C.G. Jung, *Synchronicity – An Acausal Connecting Principle* (Routledge and Kegan Paul, 1972). This essay tries to argue that events that are thought to be random, yet occur simultaneously, are connected by an additional principle that goes beyond the scientific principle of causality. Some of Jung's patients were quantum physicists (Wolfgang Pauli, one of the discoverers of quantum physics, was the most prominent one) so Jung was quite familiar with the fact that chance plays a key role in modern physics. It is interesting to see how he navigates his way between blind chance and overpowering determinism.

O. Ulfbeck and A. Bohr, Genuine fortuitousness. Where did that click come from?, *Foundations of physics*, **11**, 757 (2001). Here the authors argue that randomness has to be acknowledged as fundamental in quantum physics which means that the link between cause and effect is necessarily severed. As a natural consequence, clicks in detectors are fortuitous and cannot be attributed to the underlying existence of particles. This point of view is discussed at length in the final chapter of the book.

V. Vedral, Is reality a quantum hocus pocus?, *Straits Times*, 23 February 2008. This is the first place where I described the quantum game of cards. The game itself was introduced to me by my friend Janet Anders, a physicist at University College, London. Another similar representation of quantum physics in terms of games is to relate it to the 'game of twenty questions', as first done by Wheeler.

The idea here is for one person to imagine an object and for another to guess the object by asking 'yes/no' questions. 'Is it small?', or 'Is it material?' and so on…As the questions proceed, the guesser narrows down the range of possibilities and is, after twenty questions, in a good position to give the right answer. The quantum analogy arises by changing the game so that the first person does not imagine anything initially and then evolves an image through being consistent with his answers to the twenty questions asked. This, of course, makes it difficult for the guesser, but, if the questions are skilfully chosen, could still lead to very little choice in the end.

INDEX

Advaita Vedanta 194, 201, 214
Alice and Bob 31, 33, 38, 94, 111
analogue 18, 50, 51
Ancient Greeks 28, 32, 35, 191, 197
Anderson, Philip 96, 101
anthropic principle 55, 196
Archimedes 175, 176, 177, 186, 188,
   191, 223
Aristotle 28
Atkins, Peter 210, 220

Bacon, Roger 20
beam-splitter 119, 120, 121, 122,
   141, 157, 158, 159, 202
Bekenstein, Jacob 185, 186, 187, 224
Bell Laboratories 30, 139
Bennett, Charles ix, 73, 74, 127
Blackjack 77, 81, 82
Bohr, Niels 19, 117, 193
Boltzmann, Ludwig 62, 63, 73, 87, 93
Boole, George 32, 35, 136, 156
Boolean logic 44, 50, 112, 113, 136, 190
Brassard, Giles 127

Cairns-Smith, Alexander
   Graham 55
Calvino, Italo 15, 16, 17, 18, 19, 20, 22,
   23, 177, 205, 209, 212, 219
Cappadocian Fathers 194, 195, 214
Castells, Manuel 103
Chaitin, Gregory 167, 203, 223
Church, Alonzo 135, 151

Church–Turing thesis 135, 151
City of London 80, 97
Clausius, Rudolph 35
Copleston, Reverend 153
creation ex nihilo 5, 8, 9, 12, 18, 23, 55,
   56, 172
Crick, Francis 47, 54
cryptography 31, 123, 124, 126, 127,
   128, 132, 138, 145, 171, 222

Darwin, Charles 52
Dawkins, Richard 52, 89, 221
de Martini, Francesco 161
decoherence 145
Deutsch, David 9, 10, 41, 44, 138,
   152, 222
DNA 47, 48, 49, 51, 52, 53, 54, 55, 56, 75,
   95, 111, 132, 146, 149, 150, 191, 209

Eddington, Sir Arthur 184, 200, 201
Ehrenfest, Paul 62
Einstein, Albert 119, 129, 178, 183, 184,
   193
Ekert, Artur 127
emptiness 199, 200, 201, 203, 215, 218
entanglement 129, 160, 222
entropy 35, 37, 50, 60–73, 76, 86–90, 93,
   107, 130, 131, 147, 161, 162, 169, 173,
   179, 180–1, 189, 190, 211, 213, 224

Feuerbach, Ludwig 7, 12
Feynman, Richard 138

First Law 63, 66, 86, 184
Fleming, Graham 148, 149, 223
Franklin, Rosalind 47
Fredkin, Ed 192, 219
Freyn, Michael 117

Gabor, Denis 182
Galileo Galilei 191
Gauss, Karl Friedrich 106, 223
General relativity 178, 183, 184
Ghosh, Syantani 147
Gisin, Nicolas 128
gravitational lensing 184
Grover, Lov 139, 140, 141, 144, 148–50

Hawking, Stephen 27, 185
Heisenberg, Werner 117, 222
Hewlett Packard 46
Holographic principle 168, 181, 183,
    184, 187
Huxley, Thomas Henry 155, 220

Increasing returns, principle of 87,
    108
information theory 30–5, 79, 80, 83,
    91, 93, 94, 104, 108, 111–15, 131, 132,
    137, 138, 166, 167, 183, 190, 219–22
Intel 135, 136
Ising, Ernest 101, 102, 104, 105

Jacobson, Ted 183, 185

Kelly, Robert 78, 79, 84, 87, 221
Kepler, Johannes 197
Key Distribution Problem 127
Khayyam, Omar 23
Kolmogorov, Andrey 162, 163, 167,
    170, 177, 203

Landauer, Rolf 72, 73, 74
Landsberg, Peter 89

Laplace, Pierre Simon de 29, 113, 114
Laplace's demon 113, 114
law without law 8, 9, 23, 44, 212, 215
Leeds 26, 116, 118, 219
Leibniz, Gottfried Leibniz 8, 166, 167,
    211
logarithm 29, 61, 168

Marshall, John 26
Maxwell, James Clerk 69–73, 76, 84, 88,
    111, 147, 148, 169, 182, 213, 220, 221
Maxwell's demon 69, 72, 73, 76, 84, 88,
    111, 147, 148, 169, 182, 213, 220, 221
Mayer, Robert 62
Michelangelo 204
Moore, Gordon 135, 137, 172
Murray, Matthew 26
mutual information 93–6, 105, 108,
    128, 131, 144, 179, 180, 183, 187,
    199, 204, 211

Napier, John 29, 35
Newton, Isaac 1, 35, 113, 156, 165, 174,
    197, 198, 223
Newton's laws 1, 156
Nietzsche, Friedrich 60, 62

Occam, William of 18, 166
Onsager, Lars 101

Pauling, Linus 47
Phase transition 95, 96, 99, 101–5,
    108, 169
photon 119–22, 141, 144, 146, 157–9,
    161, 163, 201, 202, 207
Popper, Karl 164, 165, 166, 168, 174,
    177, 187, 193, 222

quantum computer 22, 137, 139–51,
    160, 173, 186, 188, 190–2, 195, 215
quantum teleportation 159, 190

quantum theory 18, 19, 21, 22, 114,
    116, 117, 118, 122, 123, 124, 127, 128,
    132, 149, 153, 155, 157, 167, 177, 178,
    180, 190, 210, 215
qubit 131, 132, 133, 137, 144, 145, 146,
    190, 215

Reeves, Keanu 197, 217
Rovelli, Carlo 167
Rumsfeld, Donald 115
Russell, Bertrand 59, 60, 153, 220

Schelling, Thomas 92, 93, 104,
    106, 221
Schrödinger, Erwin 47, 54, 68, 71,
    158, 220
Schumacher, Ben 131
scytale 124
Second Law 20, 35, 58–64, 66–9, 71,
    72, 74, 76, 85, 86, 173, 174, 192,
    215, 217, 220
Shannon, Claude 30–8, 50, 61, 68,
    76, 78–84, 86, 90, 91, 106–8,
    111–15, 125, 131, 136, 139, 161, 162,
    168, 174, 189, 190, 211, 215, 220
Shannon's information 31, 34, 79, 83,
    91, 112–14, 131, 174, 189, 190, 222
Shaw, George Bernard 58, 62
Shor, Peter 139, 142, 144
Shor's algorithm 139, 142
Singapore 146
Six Degree of Separation 97, 99
Socrates 17
Spinelli 146

Spooner, Jon 117
Spurlock, Morgan 69
Starbucks 45, 46
superposition 118, 119, 139, 141, 144,
    145, 146, 147
Susskind, Leonard 181, 182
Szilard, Leo 72

Tao Te Ching 218
Third Law of thermodynamics 75, 86
Thorp, Edward 78, 79, 81, 82, 221
Toffoli, Tom 192, 219
Tsallis, Constantino 107
Turing, Alan 135, 151, 163, 222

ultimate theory 22
Universal computer 135, 163, 166

Vegas, Las 77, 79, 82, 221
Via Negativa 194
von Neumann, John 35, 37–40, 42,
    44–8, 71, 135, 198, 199, 203, 220

Wall Street 80
Warburton, Frederic 87
Watson, James 47, 54
Wheeler, John 8, 9, 10, 41, 44, 131, 138,
    183, 217, 225
Wilkins, Maurice 47

Zeilinger, Anton 161
Zipf law 34, 103
Zipf, George 34
Zuse, Konrad 192, 219